全国高等职业教育规划教材

Android 移动应用开发案例教程

主编 范美英
参编 张晓蕾 齐京 付强 等
主审 刘瑞新

机械工业出版社

本书是一部关于 Android 开发的案例教程，采用图文并茂的方式，由浅入深、循序渐进地向读者介绍了 Android 程序设计的核心概念和技术。书中内容主要包括 Android 开发环境的配置、常见资源的使用、常用的视图组件、Android 应用程序的四大组件（Activity、BroadcastReceiver、Service、ContentProvider）的使用方法、数据存储技术、Android 应用程序中绘制 2D 图形的方法等。其中，前 7 章的最后一节均为"动手实践"，它是对各章所讲核心技术的小结，为了方便读者实践，这节中的"操作提示"还提供了操作步骤和核心技术点拨。第 8 章则为读者全方位展示了如何综合运用 Android 技术完成"数独"项目的设计与开发。

本书适合作为高职院校计算机等相关专业的教学用书，同时也适合 Android 应用开发的初学者学习使用。

本书配套授课电子课件和源代码，需要的教师可登录 www.cmpedu.com 免费注册、审核通过后下载，或联系编辑索取（QQ：1239258369，电话：010-88379739）。

图书在版编目（CIP）数据

Android 移动应用开发案例教程/范美英主编．—北京：机械工业出版社，2015.8（2019.1 重印）
全国高等职业教育规划教材
ISBN 978-7-111-50931-8

Ⅰ.①A… Ⅱ.①范… Ⅲ.①移动终端-应用程序-程序设计-高等职业教育-教材 Ⅳ.①TN929.53

中国版本图书馆 CIP 数据核字（2015）第 176781 号

机械工业出版社（北京市百万庄大街 22 号　邮政编码 100037）
策划编辑：鹿　征　　责任编辑：鹿　征
责任校对：张艳霞　　责任印制：乔　宇
三河市骏杰印刷有限公司印刷
2019 年 1 月第 1 版·第 4 次印刷
184mm×260mm·18.25 印张·451 千字
6 601-8 100 册
标准书号：ISBN 978-7-111-50931-8
定价：43.00 元

凡购本书，如有缺页、倒页、脱页，由本社发行部调换

电话服务　　　　　　　　　　　网络服务
服务咨询热线：(010)88379833　　机 工 官 网：www.cmpbook.com
　　　　　　　　　　　　　　　　机 工 官 博：weibo.com/cmp1952
读者购书热线：(010)88379649　　教育服务网：www.cmpedu.com
封面无防伪标均为盗版　　　　　　金 书 网：www.golden-book.com

全国高等职业教育规划教材计算机专业
编委会成员名单

主　　任　　周智文

副 主 任　　周岳山　林　东　王协瑞　张福强
　　　　　　陶书中　眭碧霞　龚小勇　王　泰
　　　　　　李宏达　赵佩华

委　　员　（按姓氏笔画顺序）
　　　　　　马　伟　马林艺　万雅静　万　钢
　　　　　　卫振林　王兴宝　王德年　尹敬齐
　　　　　　史宝会　宁　蒙　安　进　刘本军
　　　　　　刘剑昀　刘新强　刘瑞新　乔芃喆
　　　　　　余先锋　张洪斌　张瑞英　李　强
　　　　　　何万里　杨　莉　杨　云　贺　平
　　　　　　赵国玲　赵增敏　赵海兰　钮文良
　　　　　　胡国胜　秦学礼　贾永江　徐立新
　　　　　　唐乾林　陶　洪　顾正刚　曹　毅
　　　　　　黄能耿　黄崇本　裴有柱

秘 书 长　　胡毓坚

出 版 说 明

《国务院关于加快发展现代职业教育的决定》指出：到2020年，形成适应发展需求、产教深度融合、中职高职衔接、职业教育与普通教育相互沟通，体现终身教育理念，具有中国特色、世界水平的现代职业教育体系，推进人才培养模式创新，坚持校企合作、工学结合，强化教学、学习、实训相融合的教育教学活动，推行项目教学、案例教学、工作过程导向教学等教学模式，引导社会力量参与教学过程，共同开发课程和教材等教育资源。机械工业出版社组织全国60余所职业院校（其中大部分是示范性院校和骨干院校）的骨干教师共同策划、编写并出版的"全国高等职业教育规划教材"系列丛书，已历经十余年的积淀和发展，今后将更加紧密结合国家职业教育文件精神，致力于建设符合现代职业教育教学需求的教材体系，打造充分适应现代职业教育教学模式的、体现工学结合特点的新型精品化教材。

"全国高等职业教育规划教材"涵盖计算机、电子和机电三个专业，目前在销教材300余种，其中"十五""十一五""十二五"累计获奖教材60余种，更有4种获得国家级精品教材。该系列教材依托于高职高专计算机、电子、机电三个专业编委会，充分体现职业院校教学改革和课程改革的需要，其内容和质量颇受授课教师的认可。

在系列教材策划和编写的过程中，主编院校通过编委会平台充分调研相关院校的专业课程体系，认真讨论课程教学大纲，积极听取相关专家意见，并融合教学中的实践经验，吸收职业教育改革成果，寻求企业合作，针对不同的课程性质采取差异化的编写策略。其中，核心基础课程的教材在保持扎实的理论基础的同时，增加实训和习题以及相关的多媒体配套资源；实践性较强的课程则强调理论与实训紧密结合，采用理实一体的编写模式；涉及实用技术的课程则在教材中引入了最新的知识、技术、工艺和方法，同时重视企业参与，吸纳来自企业的真实案例。此外，根据实际教学的需要对部分课程进行了整合和优化。

归纳起来，本系列教材具有以下特点：

1) 围绕培养学生的职业技能这条主线来设计教材的结构、内容和形式。

2) 合理安排基础知识和实践知识的比例。基础知识以"必需、够用"为度，强调专业技术应用能力的训练，适当增加实训环节。

3) 符合高职学生的学习特点和认知规律。对基本理论和方法的论述容易理解、清晰简洁，多用图表来表达信息；增加相关技术在生产中的应用实例，引导学生主动学习。

4) 教材内容紧随技术和经济的发展而更新，及时将新知识、新技术、新工艺和新案例等引入教材。同时注重吸收最新的教学理念，并积极支持新专业的教材建设。

5) 注重立体化教材建设。通过主教材、电子教案、配套素材光盘、实训指导和习题及解答等教学资源的有机结合，提高教学服务水平，为高素质技能型人才的培养创造良好的条件。

由于我国高等职业教育改革和发展的速度很快，加之我们的水平和经验有限，因此在教材的编写和出版过程中难免出现问题和疏漏。我们恳请使用这套教材的师生及时向我们反馈质量信息，以利于我们今后不断提高教材的出版质量，为广大师生提供更多、更适用的教材。

机械工业出版社

前　言

　　Android 及它的绿色小机器人标志和苹果 iPhone 一样风靡世界，掀起了移动领域最具影响力的风暴。从本质上来看，Android 是一个以 Linux 为基础的开源移动设备操作系统，如今它主要被用于智能手机、平板电脑等移动设备。它一直由 Google 成立的 Open Handset Alliance（OHA，开放手持设备联盟）领导并持续开发。Android 目前已发布的最新版本为 Android 5.0（Lollipop）。

　　本书以 Android 4.2 为开发平台，使用 Eclipse 开发环境，以 Java 为开发语言，比较完整地介绍了开发 Android 移动应用所需要的知识和技术。本书的主要内容包括配置 Android 开发环境的方法，Android 中的常见资源，Android 中的基本视图组件与高级视图组件，Android 的四大组件（Activity、BroadcastReceiver、Service、ContentProvider），SQLite 数据库存取技术，SharedPreferences 的定义与使用，Android 中对音频、视频等各种媒体的使用与处理技术，以及综合实训项目"快乐数独"等内容。每章均有典型的演练和练习题目，以帮助教师演示和学生练习。

　　在学习本课程前，要求学生具有基本的界面设计常识和 Java 程序设计语言基础，能够熟练使用 Eclipse 开发环境。

　　本书的特点是图文并茂，案例的设计和实现过程详细完整，份量适中，内容切割分明且完备，教材中的每个知识点都相对独立，为学生随时开始学习提供了可能。通过多次学习和练习书中的各个案例，学生可以不断积累界面设计的经验，掌握界面设计的要领，逐步领会用户体验的重要性，了解 Android UI 的规范和核心原则，深入理解 MVC（Model – View – Controller，模型 – 视图 – 控制器）的概念模式，有利于养成良好的代码编程风格。本书在编写风格上，力求深入浅出，尽量将知识融于浅显、常见的案例之中，以便学习者可以轻松地学到知识。

　　在教学中我们发现，学生在学习移动应用开发时，往往知其然不知其所以然，以至于不能举一反三。为此，我们在本书中从学生喜闻乐见的案例操作开始，在完成案例练习后将涉及到的知识和相关的背景知识进行了必要的介绍。这对于老师教学和学生复习巩固起到了必要的补充作用。

　　本书的每章均有动手实践环节，这一部分中的"操作提示"将为读者提供操作步骤和核心技术点拨，为课后独立实践提供了保障；同时，动手实践项目也是对各章所讲核心技术的小结。在知识内容的细节介绍上，采用了符合认知规律的形式，即先引出概念，再介绍语法格式，然后介绍方法步骤，最后结合案例给出使用场景和方式。

　　本书的另一个特点是合理取舍，因为受到课时的限制，课堂上没有过多的时间讲授全部内容，本书选取 Android 中的最基本的知识来介绍，未对复杂不常用的知识进行介绍，如传感器、手机服务等。这些内容完全可以在掌握了基础知识之后，随着经验的积累和实践中项目开发的需要，通过查询 Google 提供的 API（Application Programming Interface，应用程序编程接口）文档等掌握。

本书的主要作者是具有丰富教学经验的教师与经验丰富的企业开发工程师，优势互补保障了本书的质量，使得本书更贴近实际，是校企结合的结晶、范例。主审刘瑞新更是拥有丰富的教材编写经验，对案例细节的取舍进行了严格把关，使各个案例均符合教学使用的需要。

本书由范美英主编，张晓蕾、齐京、付强等编著，编写作者有范美英（第1、4、5、6、8章），张晓蕾（第2章），齐京（第3章），付强（第7章），本书课件的制作由王鹏、李成、李薇、张娟、田新莲、王彦峰、郭林、徐晓楠、梁广海等完成，教材中的许多代码都经由徐丽、乔宇青、呼昊、刘晓星等人编码并提供技术支持，全书由范美英主编、统稿，刘瑞新审核。由于编著者水平有限，书中错误与疏漏之处在所难免，敬请读者批评指正。

编　者

目 录

出版说明
前言
第1章 Android 概述 ········· 1
　1.1　搭建 Android 开发环境 ········· 1
　　1.1.1　安装 Java ········· 1
　　1.1.2　安装 Eclipse ········· 1
　　1.1.3　安装 Android SDK ········· 2
　　1.1.4　配置 Eclipse ········· 3
　　1.1.5　高效的 Eclipse ADT Bundle ········· 5
　1.2　Android 虚拟设备（AVD） ········· 5
　　1.2.1　AVD 概述 ········· 5
　　1.2.2　创建 AVD ········· 6
　　1.2.3　模拟器与真机 ········· 8
　1.3　Android 中的常用工具 ········· 8
　　1.3.1　ADB 的使用 ········· 8
　　1.3.2　DDMS 的使用 ········· 9
　1.4　Android 系统架构 ········· 11
　　1.4.1　Linux 内核 ········· 12
　　1.4.2　库 ········· 12
　　1.4.3　Android 运行时 ········· 13
　　1.4.4　应用程序框架 ········· 13
　　1.4.5　应用程序 ········· 14
　1.5　Android 平台简介 ········· 14
　　1.5.1　Android 平台的特性 ········· 14
　　1.5.2　Android SDK 版本的特点 ········· 15
　1.6　实例1："你好，Android" ········· 18
　　1.6.1　创建应用程序 ········· 18
　　1.6.2　在模拟器上运行应用程序 ········· 20
　　1.6.3　Android 应用程序的项目结构 ········· 21
　1.7　动手实践1：第1个 Android 应用 ········· 23
　　1.7.1　功能要求 ········· 23
　　1.7.2　操作提示 ········· 23
第2章　Android 中的资源 ········· 25
　2.1　实例1：千变万化背景色 ········· 25

  2.1.1 功能要求与操作步骤 ·················· 25
  2.1.2 颜色（color）资源的定义和使用 ··········· 28
 2.2 实例2：屏蔽身份证部分信息 ················ 29
  2.2.1 功能要求与操作步骤 ·················· 29
  2.2.2 字符串（string）资源的定义与使用 ········ 34
 2.3 实例3：渐现"四书五经" ················· 35
  2.3.1 功能要求与操作步骤 ·················· 35
  2.3.2 线性布局（LinearLayout）的定义与使用 ····· 39
 2.4 实例4：初读"大学" ···················· 40
  2.4.1 功能要求与操作步骤 ·················· 40
  2.4.2 帧布局（FrameLayout）的定义与使用 ······· 42
 2.5 实例5：办公电话一览 ···················· 43
  2.5.1 功能要求与操作步骤 ·················· 43
  2.5.2 表格布局（TableLayout）的定义与使用 ····· 46
 2.6 实例6：梅花效果首界面 ·················· 46
  2.6.1 功能要求与操作步骤 ·················· 46
  2.6.2 相对布局（RelativeLayout）的定义与使用 ··· 50
 2.7 实例7：DIY计算器 ····················· 51
  2.7.1 功能要求与操作步骤 ·················· 51
  2.7.2 网格布局（GridLayout）的定义与使用 ······ 54
 2.8 实例8：美食背后的故事 ·················· 55
  2.8.1 功能要求与操作步骤 ·················· 55
  2.8.2 布局的嵌套使用 ····················· 59
 2.9 实例9：简易文本阅读器 ·················· 60
  2.9.1 功能要求与操作步骤 ·················· 60
  2.9.2 菜单（Menu）资源的定义与使用 ··········· 66
 2.10 动手实践2：紫禁城一日游 ················· 67
  2.10.1 功能要求 ························· 67
  2.10.2 操作提示 ························· 68

第3章 Android中的基本视图组件 ················· 70
 3.1 实例1：新闻摘要与详情 ·················· 70
  3.1.1 功能要求与操作步骤 ·················· 70
  3.1.2 文本显示组件（TextView）的定义与使用 ···· 74
 3.2 实例2：微信登录 ······················ 75
  3.2.1 功能要求与操作步骤 ·················· 75
  3.2.2 编辑框（EditText）的定义与使用 ········· 78
  3.2.3 按钮（Button）的定义与使用 ············ 79
  3.2.4 信息提示框（Toast）使用简介 ············ 80
 3.3 实例3：注册应用账号 ··················· 80

3.3.1 功能要求与操作步骤 ……………………………………………………… 80
3.3.2 下拉列表框（Spinner）的定义与使用 …………………………………… 86
3.3.3 复选框（CheckBox）的定义与使用 ……………………………………… 87
3.3.4 图像按钮（ImageButton）的定义与使用 ………………………………… 88
3.3.5 短信管理器（SmsManager）使用简介 …………………………………… 89
3.4 实例4：完善个人资料 ……………………………………………………………… 90
3.4.1 功能要求与操作步骤 ……………………………………………………… 90
3.4.2 单选按钮组（RadioGroup与RadioButton）的定义与使用 ……………… 99
3.4.3 图像框（ImageView）的定义与使用 ……………………………………… 100
3.4.4 警告对话框（AlertDialog与AlertDialog.Builder）使用简介 …………… 101
3.5 动手实践3：为友秀宝 ……………………………………………………………… 103
3.5.1 功能要求 …………………………………………………………………… 103
3.5.2 操作提示 …………………………………………………………………… 104

第4章 Android中的高级视图组件 108

4.1 实例1：随心换肤 …………………………………………………………………… 108
4.1.1 功能要求与操作步骤 ……………………………………………………… 108
4.1.2 图片切换器（ImageSwitcher）的定义与使用 …………………………… 113
4.1.3 文本切换器（TextSwitcher）的定义与使用 …………………………… 115
4.1.4 设置手机桌面背景简介 …………………………………………………… 116
4.2 实例2：居家助手 …………………………………………………………………… 116
4.2.1 功能要求与操作步骤 ……………………………………………………… 116
4.2.2 选项卡（TabHost）的生成与使用 ………………………………………… 123
4.2.3 标签（TabSpec）的定义与使用 …………………………………………… 125
4.2.4 日期选择器（DatePicker）与时间选择器（TimePicker） ……………… 126
4.2.5 显示地图的基本步骤 ……………………………………………………… 127
4.2.6 TabActivity的取代者FragmentActivity …………………………………… 128
4.3 实例3：全球名校快查 ……………………………………………………………… 132
4.3.1 功能要求与操作步骤 ……………………………………………………… 132
4.3.2 自动完成文本框（AutoCompleteTextView）的定义与使用 …………… 136
4.3.3 列表视图（ListView）的定义与使用 ……………………………………… 138
4.4 动手实践4：休闲时分 ……………………………………………………………… 139
4.4.1 功能要求 …………………………………………………………………… 139
4.4.2 操作提示 …………………………………………………………………… 140

第5章 Android应用程序的组成 148

5.1 实例1：身体质量指数测试 ………………………………………………………… 148
5.1.1 功能要求与操作步骤 ……………………………………………………… 148
5.1.2 活动（Activity）的定义与使用 …………………………………………… 153
5.1.3 意图（Intent）的定义与使用 ……………………………………………… 156
5.2 实例2：编辑商品信息 ……………………………………………………………… 159

　　　　5.2.1　功能要求与操作步骤 …………………………………………………… 159
　　　　5.2.2　获取 Activity 返回值的方法 ……………………………………………… 166
　　5.3　实例3：快速联系 ……………………………………………………………… 167
　　　　5.3.1　功能要求与操作步骤 …………………………………………………… 167
　　　　5.3.2　调用拨号程序和短信程序的方法 ………………………………………… 170
　　　　5.3.3　ContentProvider 共享数据的方法 ………………………………………… 171
　　5.4　实例4：闹钟服务 ……………………………………………………………… 172
　　　　5.4.1　功能要求与操作步骤 …………………………………………………… 172
　　　　5.4.2　服务（Service）的定义 ………………………………………………… 177
　　　　5.4.3　广播接收器（BroadcastReceiver）的定义与使用 ………………………… 178
　　　　5.4.4　四大组件之间的关系 …………………………………………………… 179
　　5.5　动手实践5：掌上电子邮件 …………………………………………………… 179
　　　　5.5.1　功能要求 ………………………………………………………………… 179
　　　　5.5.2　操作提示 ………………………………………………………………… 180

第6章　Android 中的数据存取　181
　　6.1　实例1：保存偏好设置 ………………………………………………………… 181
　　　　6.1.1　功能要求与操作步骤 …………………………………………………… 181
　　　　6.1.2　SharedPreferences 的定义与使用 ………………………………………… 188
　　6.2　实例2：贴身账簿 ……………………………………………………………… 189
　　　　6.2.1　功能要求与操作步骤 …………………………………………………… 189
　　　　6.2.2　文件（File）的定义与使用 ……………………………………………… 195
　　6.3　实例3：备忘随行 ……………………………………………………………… 197
　　　　6.3.1　功能要求与操作步骤 …………………………………………………… 197
　　　　6.3.2　SQLite 数据库的基本使用方法 …………………………………………… 207
　　6.4　动手实践6：查账单 …………………………………………………………… 209
　　　　6.4.1　功能要求 ………………………………………………………………… 209
　　　　6.4.2　操作提示 ………………………………………………………………… 209

第7章　Android 中的媒体处理　211
　　7.1　实例1：绘制五星红旗 ………………………………………………………… 211
　　　　7.1.1　功能要求与操作步骤 …………………………………………………… 211
　　　　7.1.2　常用的绘图类 …………………………………………………………… 215
　　　　7.1.3　绘制简单图形的基本方法 ……………………………………………… 217
　　7.2　实例2：放大镜看 SD 卡中的图 ……………………………………………… 219
　　　　7.2.1　功能要求与操作步骤 …………………………………………………… 219
　　　　7.2.2　访问图库中的图像 ……………………………………………………… 224
　　7.3　实例3：扣篮瞬间 ……………………………………………………………… 226
　　　　7.3.1　功能要求与操作步骤 …………………………………………………… 226
　　　　7.3.2　Android 动画技术简介 …………………………………………………… 229
　　7.4　实例4：悦视播放器 …………………………………………………………… 232

7.4.1　功能要求与操作步骤 ··· 232
　　　7.4.2　音频的播放与录制 ··· 245
　　　7.4.3　使用SurfaceView播放视频的步骤 ································· 247
　7.5　动手实践7：迷你画板 ·· 248
　　　7.5.1　功能要求 ·· 248
　　　7.5.2　操作提示 ·· 249
第8章　综合实训：快乐数独 ··· 257
　8.1　数独（Sudoku）简介 ·· 257
　　　8.1.1　数独概述 ·· 257
　　　8.1.2　数独的游戏规则与技巧 ·· 257
　8.2　项目功能分析 ·· 258
　　　8.2.1　项目的主要功能 ··· 258
　　　8.2.2　"自定义设置"菜单 ·· 258
　　　8.2.3　"软键盘"与"提示" ·· 259
　8.3　准备所需资源 ·· 259
　　　8.3.1　图片（res/drawable-x/）·· 259
　　　8.3.2　音频（res/raw/）··· 260
　　　8.3.3　数组（res/values/arrays.xml）······································ 260
　　　8.3.4　颜色（res/values/colors.xml）······································ 260
　　　8.3.5　字符串（res/values/strings.xml）·································· 260
　　　8.3.6　动画（res/anim/cycle.xml和shake.xml）······················· 261
　8.4　界面设计 ·· 261
　　　8.4.1　首界面（res/layout/activity_main.xml）························· 261
　　　8.4.2　游戏界面（SudokuView.java）····································· 262
　　　8.4.3　设置界面（res/xml/settings.xml）································· 266
　　　8.4.4　软键盘界面（res/layout/keypad.xml）·························· 267
　　　8.4.5　菜单界面（res/menu/menu.xml）································· 267
　8.5　数据库设计 ··· 267
　　　8.5.1　定义数据库常量类（Constants.java）···························· 267
　　　8.5.2　定义数据库辅助类（DBHelper.java）···························· 268
　8.6　功能实现与完善 ··· 269
　　　8.6.1　首界面中按钮与菜单的功能（MainActivity.java）············ 269
　　　8.6.2　"设置"的实现 ·· 272
　　　8.6.3　"新游戏"与"继续"功能（NewGameActivity.java）··········· 273
　　　8.6.4　"关于"功能（AboutActivity.java）······························· 279
　　　8.6.5　修改配置文件（AndroidManifest.xml）························· 279
　　　8.6.6　项目的完善方向 ··· 280

第1章 Android 概述

Android 一词的中文释义是"机器人",它是 Google 公司于 2007 年 11 月宣布的基于 Linux 平台开发的一款手机操作系统。简单地说,Android 是基于 Java 并运行在 Linux 内核上的操作系统,是全球第一个完整的、完全开放的手机平台。

本章介绍如何搭建 Android 开发环境、如何创建和使用 AVD、Android 应用程序的文件结构、Android 中的常用工具、Android 平台概况,最后演示如何创建一个可运行的应用程序:"你好,Android!"。

1.1 搭建 Android 开发环境

Android SDK(软件开发包)可以在 Windows、Linux 或者 Mac OS X 上运行,建议在开发前,选择一款自己最熟悉的操作系统作为平台。本书以 Windows 为例介绍如何搭建 Android 开发环境。

1.1.1 安装 Java

Android 是基于 Java 并运行在 Linux 内核上的一款手机操作系统,所以首先要安装 Java,并且需要使用 JDK 5 或者 JDK 6。

在 Windows 上安装 JDK 时,首先从 Oracle 官方网站(http://java.oracle.com/technetwork/java/index.html)下载最新的 JDK 版本。双击下载得到的 exe 安装文件后即可打开安装向导,在接受许可、选择好需要安装的组件和安装路径后,向导即可自动完成安装。

在安装完 JDK 后,需要将环境变量 Path 中加入 JDK 文件夹下 bin 的路径。现假设将 JDK 安装在 D:\JDK 中,设置环境变量的步骤为:右击"我的电脑",在右键菜单中单击"属性"命令,在弹出的对话框中选择"高级"选项卡。单击"环境变量"按钮,在弹出的对话框中选择"系统变量"中的"Path",然后单击"编辑"按钮,在"变量值"的结束处添加"D:\JDK\bin"。

要确定已安装的版本是否正确,可以单击"开始"菜单,选择"运行"选项,在出现的对话框中输入"cmd"命令,打开命令行窗口,并在其中运行"java – version"命令。如果显示类似下面的信息,版本是"1.6.其他数字"或更高,则表示 JDK 安装成功。

```
C:\Users\hp > java – version
java version "1.6.0_20"
Java(TM) SE Runtime Environment (build 1.6.0_20 – b02)
Java HotSpot(TM) Client VM (build 16.3 – b01,mixed mode,sharing)
```

1.1.2 安装 Eclipse

安装好 JDK 后,如果计算机上没有 Java 开发环境,则需要安装 Eclipse 或其他 IDE(Integrated Development Environment，集成开发环境）。在 Eclipse 的官方网站（http://www.eclipse.org/downloads）下载到最新或熟悉的 Eclipse IDE for Java Developers 可用版本的压缩文件（本书中使用 Eclipse Helios），然后将之解压到合适的目录下。进入解压后的目录，用户可以看到名为"eclipse.exe"的可执行文件，双击此文件即可运行 Eclipse。如果安装后是第一次启动 Eclipse，用户将会看到选择工作空间的提示，此时根据需要选择即可。

1.1.3 安装 Android SDK

安装好 JDK 和 Eclipse 后，需要安装 Android SDK。从 Android 2.0 开始，Android SDK 被分为两部分：SDK 启动程序软件包和 SDK 组件。

首先从 Android 下载页（http://d.android.com/sdk）上获得合适的.zip 软件包（本教材中使用 Android SDK 4.2），然后将此文件解压到合适的目录下，此目录即为 SDK 的安装目录。另外，还需将 SDK 的 tools 目录添加到 Path 环境变量中。此处假设将 Android SDK 解压到了 D:\AndroidSDK 目录下，解压完成后，需要将 Path 环境变量的值结尾处添加"; D:\AndroidSDK\tools"。设置 Path 环境变量的方法与 JDK 处一致，此处不再详述。在设置好环境变量后，如果在 DOS 命令窗口中输入"android -h"看到 Android 的帮助信息，则说明 Android SDK 安装成功。

接下来，在安装目录下找到 SDK 组件的管理程序"SDK Manager.exe"，在接入互联网的前提下运行该程序，选择"Available packages"选项，在每个需要的软件包旁边都添加复选标记，并单击"Install Selected"，效果如图 1-1 所示。

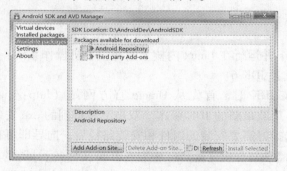

图 1-1　SDK Manager 运行效果图

安装程序会显示一个可用组件列表，包括各种文档、平台、附加软件库等，选择"Accept All"，然后单击"Install"，结果如图 1-2 所示。这样所有列出的组件都将被下载并安装到 SDK 的安装目录中。为了提高速度，可以单独接受或拒绝各个组件，而不必一次全部安装。

图 1-2　SDK 组件列表示意图

1.1.4 配置 Eclipse

为了让开发过程更轻松，Google 还编写了一个称为 ADT（Android Development Tools，安卓软件开发工具）的 Eclipse 插件，在 Eclipse Helios 中配置 ADT 的参考步骤如下。

1) 运行"Edipse"，打开"Help"菜单，在弹出的对话框中单击"Install New Software"命令，如图 1-3 所示。

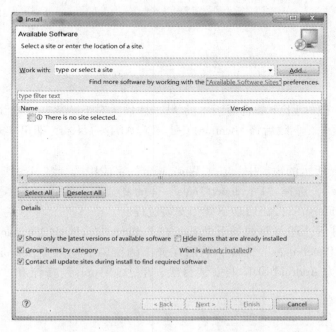

图 1-3 Install New Software 窗口

2) 在图 1-3 所示的对话框中单击"Available Software Sites"链接，弹出如图 1-4 所示的窗口。

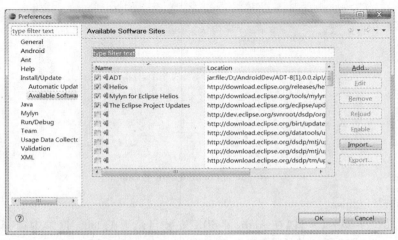

图 1-4 Available Software Sites 窗口

3）在图1-4所示的对话框中单击"Add…"按钮，弹出"Add Site"对话框。其中，Name值可以自定义，Location值处输入"http://dl-ssl.google.com/Android/eclipse/"，如图1-5所示，单击"OK"按钮。

图1-5 Add Site对话框

此处也可以从互联网上下载合适版本的ADT插件。对于SDK 4.2来说，需要21.0.0或以上版本的ADT插件。下载后将Location值处输入ADT的目录，即可实现从本地安装，这样可以更加快速。

4）回到图1-3所示的窗口中，在"Work with"字段中选择"MyADT"，列表下方将出现"Developer Tools"，选择"Developer Tools"旁的复选框，然后单击"Next"按钮，在接受许可后，单击"Finish"按钮启动下载和安装过程。

5）安装好插件后，还需要做如下配置。在Eclipse中，选择"Window"菜单，在弹出的对话框中选择"Preferences"命令，然后在弹出的窗口中选中"Android"选项，在右侧设定SDK Location为Android SDK的安装目录，单击"Apply"按钮和"OK"按钮完成配置，如图1-6所示。

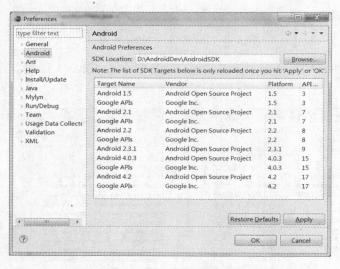

图1-6 设置Android SDK Location

6）重新启动Eclipse。

经过上述几个步骤后，可以在Android SDK安装目录下的tools目录中运行"android list targets"命令，如果能看到类似如图1-7所示的Android SDK版本列表，则表示Android的开发环境搭建完成。

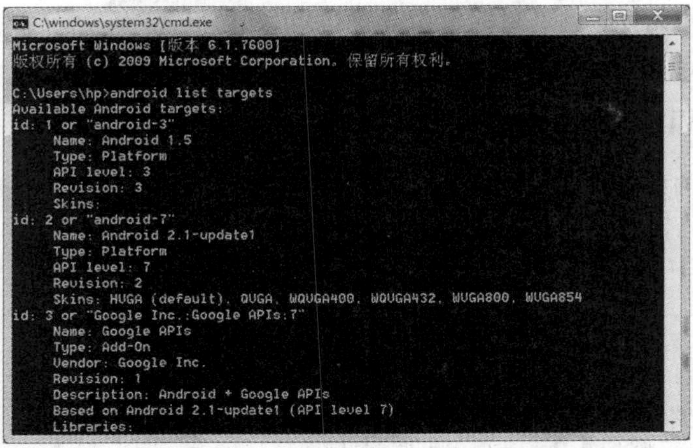

图1-7 Android SDK 版本列表

1.1.5 高效的 Eclipse ADT Bundle

随着 Android 的不断普及和发展，Android 开发环境的配置也越来越方便。Google 在 http://developer.android.com/sdk/index.html 上提供了一个新的开发工具：ADT Bundle。使用步骤如下：

1）下载 ADT Bundle 安装包，它包含了开始开发应用所需的所有组件，如 Eclipse + ADT 插件、Android SDK 工具、Android Platform – tools、最新的 Android 开发平台版本和模拟器。

2）解压下载的安装包，然后保存在一个合适的主目录下。需要注意的是不要移动 adt – bundle –< os_platform > 目录下的任何文件和目录，否则就需要手动更新 ADT。

1.2 Android 虚拟设备（AVD）

任何应用程序只有在不断地调试之后才能确定能否正常运行。在开发 Android 手机应用程序时，如何去调试呢？只要准备一台安装 Android SDK 和 Android 虚拟设备的计算机就可以了。

1.2.1 AVD 概述

AVD（Android Virtual Device）即 Android 虚拟设备，每个 AVD 模拟了一套虚拟设备来运行 Android 平台。这个平台包含独立的内核、系统图像和数据分区，还可以拥有自己的 SD 卡和用户数据等。只有在创建 AVD 之后，才能正确地启动 Android 模拟器。所谓模拟器就是指在计算机上模拟 Android 系统，然后使用该系统来调试并运行开发好的 Android 应用程序。在这个过程中，开发人员只需要利用 AVD 即可创建不同 Android 版本的模拟器，以便模拟运行一个手机操作系统。

对于 Android 程序的开发者来说，无论是在 Windows 下还是在 Linux 下，或者在 Mac OS X 下都可以顺利运行 Android 模拟器。在需要时，可以从 Android 官方网站（http://developer. android. com）免费下载单独的模拟器，也可以在 Android 开发环境下创建自己的模拟器。

1.2.2 创建 AVD

创建 AVD 有两种方式：一种是在命令窗口中创建，另外一种是借助 Eclipse IDE 的可视化窗口创建。

1. 在命令窗口中创建 AVD

打开命令窗口，可以参考如下命令行中的参数，按照"android create avd – name < AVD 的名称 > – – target < targetID >"格式创建 AVD。

```
C:\Users\hp > android createavd – – name testAVD – – target 2
Android 2.1 – update1 is a basic Android platform.
Do you wish to create a custom hardware profile [no]n
Created AVD 'testAVD'based on Android 2.1 – update1,
with the following hardware config:
hw. lcd. density = 160
```

使用上面的命令可以创建名为"testAVD"的虚拟设备，然后，在运行应用程序时，只要在 Eclipse 的 Run Configuration 中指定 AVD 的名字即可。

需要注意的是，在上面的命令执行后，系统提示是否定义一个硬件配置，即"Android 2.1 – update1 is a basic Android platform. Do you wish to create a custom hardware profile [no]"，默认选择"no"，这样可以快速地创建一个 AVD 配置。如果选择了"yes"，则需要对各种硬件参数进行设置。

AVD 的保存位置会根据一个叫做"ANDROID_SDK_HOME"的环境变量来指定，如果没有指定该变量，它会在系统默认的目录下创建。在 Windows 中的默认路径是 C:\users\<user>\.android\。在这个目录下，有一个 avd 目录，该目录下存放有创建好的 AVD 配置文件。

2. 在 Eclipse IDE 中创建 AVD

在 Eclipse IDE 的可视化环境中，选择"Window"菜单，单击"Android Virtual Device manager"命令可以弹出如图 1-8 所示的窗口。

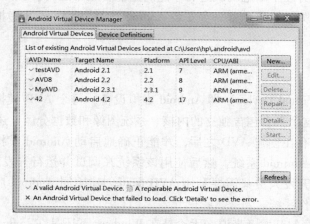

图 1-8 AVD Manager 窗口

选择左侧列表中的"Android Virtual Devices"选项，单击右侧的"New"按钮，即可弹出创建 AVD 的对话框，如图 1-9 所示，其中 Name 的值可自定义，"Targets"需要从下拉列表框中选择，SD 卡的大小以及外观特征都可以根据需要进行设置。

图 1-9 创建 AVD 示意图

创建好 AVD 后，在图 1-8 所示的 AVD 列表中，选中打算启动的模拟器，单击右侧的"Start"按钮后可弹出如图 1-10 所示的"Launch Options"窗口。

在图 1-10 所示的窗口中单击"Launch"按钮即可启动所选模拟器，基于 Android SDK 4.2 版本的模拟器效果如图 1-11 所示。

图 1-10 "Launch Options"窗口　　图 1-11 Android SDK 4.2 版本的模拟器

需要注意的是，模拟器大小不一致时，应用程序的外观会受到不同程度的影响。本书的应用程序在测试时使用的模拟器窗口大小为 320×480 像素。

1.2.3 模拟器与真机

AVD 配置的 Android 模拟器与 Android 真机很相似，但却不能完全代替真机。例如，在真机中用户可以呼叫和接听实际来电，但在模拟器中只能通过控制台模拟电话呼叫；真机可以进行视频和音频捕捉，但模拟器不支持；真机可以连接扩展耳机、蓝牙等外部设备，但模拟器不可以；真机能够确定电池电量水平、充电状态，但模拟器无法确定等。

就像真机开机一样，启动模拟器需要花费一些时间，关闭模拟器就像是真机关闭手机或取出手机电池一样，所以如果需要在模拟器上不断地调试应用程序，那么不要关闭它，以免再次启动花费不必要的转换时间。这就意味着，只要 Eclipse 在运行，就保持模拟器窗口处于打开状态，下次启动 Android 程序时，Eclipse 会注意到模拟器已经启动，它随即会向模拟器发送要运行的新程序。

开发期间，在真机上运行 Android 程序与在模拟器上运行该程序的效果几乎相同，只要将手机与计算机用 USB 连接起来并安装好手机驱动即可。只是在真机上运行时，如果模拟器窗口已打开，请将它关闭，这样只要将手机与计算机相连，应用程序就会在手机上加载并运行。

使用模拟器时，可能会遇到计算机提示系统盘空间不足之类的错误信息。这是因为 Android 模拟器每次运行时会临时生成几个扩展名为 "tmp" 的临时文件，一段时间后，这些文件会占用大量的磁盘空间，所以，出现类似问题时清理一下临时文件即可。

由于 keystore 到期也会导致应用程序不能够在模拟器中调试，此时只要将 C:\Users\<用户名>\.android 目录下的 debug.keystore 删掉，重新运行应用程序即可。

1.3 Android 中的常用工具

在 Android SDK 安装目录下有一个 tools 目录，该目录中包含了 Android 中的常用工具。另外，在 platform – tools 目录下也有一些工具。

1.3.1 ADB 的使用

ADB（Android Debug Bridge，安卓调试桥）是管理模拟器的一个通用工具，该工具的功能很多。例如，将系统文件复制到设备、从设备复制文件到系统、安装 APK 项目、查看当前设备等。

1. 查询当前模拟器实例数量

有时 adb.exe 文件放在 Android SDK 的安装目录下的 tools 目录中，有时也会存放在 platform – tools 中。在命令窗口中定位到 adb.exe 文件所在目录，即可使用 adb devices 命令来查看当前运行的 Android 模拟器实例有哪些，运行效果如图 1-12 所示。

2. 本地机器和模拟器之间相互复制文件

在开发应用程序的过程中，有时需要将模拟器中的数据文件在本地机器备份，或者将本地文件复制到模拟器中，此时就可以使用"adb pull"命令或"adb push"命令。例如，将 D:\music.mp3 文件复制到模拟器的 SD 卡中，就可以使用如下命令：

图 1-12　adb 的使用

> D:\AndroidDev\AndroidSDK\platform-tools > adb push d:\music.mp3 /sdcard/

当需要将模拟器中的文件 test.txt 复制到本地机器中的 D 盘根目录下时，可以参考如下命令。

> D:\AndroidDev\AndroidSDK\platform-tools > adb pull /sdcard/test.txt d:/

上述命令执行后，如果文件复制成功，系统会给一个类似"134 KB/s（1106708 bytes in 8.040s）"的数据传输速率的提示。

3. 安装 APK 应用程序

在互联网上可以下载到很多 Android 应用的 apk 安装程序，使用"adb install"命令即可安装一个 apk 应用到设备。例如，下面的命令即可将 D:\test.apk 安装到模拟器中。

> D:\AndroidDev\AndroidSDK\platform-tools > adb install d:\test.apk

上述命令执行后，如果程序安装成功，系统会给类似如下信息的提示。

> 1183 KB/s（1293370 bytes in 1.067s）
> 　　　pkg:/data/local/tmp/test.apk
> Success

4. 使用 shell 命令

使用 adb shell 命令可以进入 shell 命令行，此时即可运行模拟器文件系统中的/system/bin/目录下的所有命令。下面的命令即可查看当前日期，并退出 shell 命令行。

> D:\AndroidDev\AndroidSDK\platform-tools > adb shell
> # date
> date
> Wed Jan 30 13:24:38 GMT 2013
> # exit
> exit

1.3.2　DDMS 的使用

DDMS（Dalvik Debug Monitor Service，Dalvik 调试监控服务），是一个可视化的调试监控工具。在 Eclipse 中，打开"Window"菜单，单击"Open Perspective"命令，在弹出的子菜

单中选择"Other"命令。然后,在其弹出的选项卡中选中"DDMS"即可看到如图 1-13 所示的 DDMS 窗口。

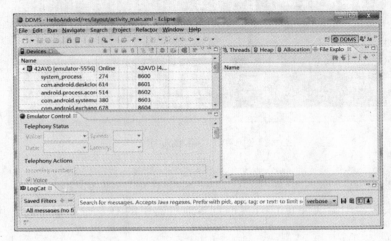

图 1-13　DDMS 窗口

在这个窗口中,可以对系统运行后台日志、系统线程、模拟器状态等进行监控。另外,该窗口还可以模拟发送短信、拨打电话、发送 GPS 位置信息,查看 sdcard 中的文件信息等。

1. 模拟拨打语音电话

在 DDMS 中选中一个模拟器,然后在如图 1-14 所示的 Emulator Control 窗口中,将"Incoming number"字段中填写电话号,如"123456",选中"Voice"单选按钮,单击"Call"按钮,即可向模拟器拨打语音电话。

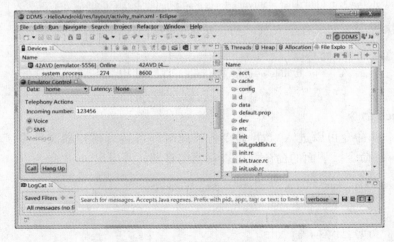

图 1-14　Emulator Control 窗口

此时的模拟器如图 1-15 所示。

2. 查看文件信息

在使用前文所提的 adb push 命令将本地机器中的 test.txt 文件复制到模拟器的 sdcard 后,即可在 DDMS 窗口中的"File Explorer"标签的"mnt"→"sdcard"下看到新增的文件,如图 1-16 所示。

 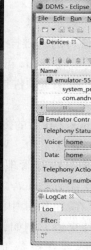 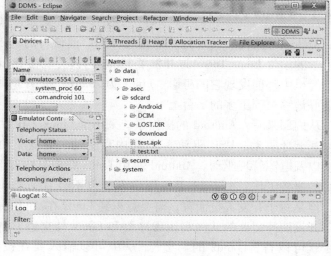

图 1-15 模拟语音电话　　　　　　　图 1-16 File Explorer 查看文件

1.4　Android 系统架构

要编写出功能良好的 Android 应用程序，需要了解组成 Android 开源软件栈的关键层和组件有哪些，这就是 Android 的总体系统架构中的内容。

Android 的系统架构如图 1-17 所示。从图中可看出 Android 系统架构分为 4 层，从下到上分别是 Linux Kernel（Linux 内核）、Libraries and Android Runtime（核心类库，含 Android 运行时环境）、Application Framework（应用程序框架）和 Applications（应用程序），并且图中的每一层都使用其下面各层所提供的服务。

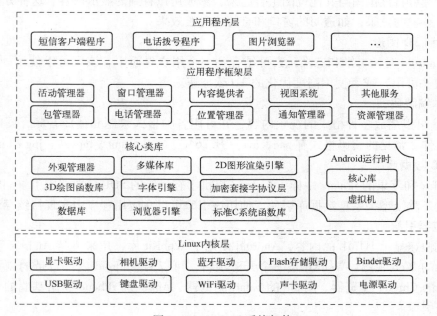

图 1-17　Android 系统架构

1.4.1 Linux 内核

Android 构建在 Linux 这个稳定且得到广泛认可的基础操作系统之上。1991 年，赫尔辛基大学的学生 Linus Torvalds 开发了 Linux 内核。现在 Linux 可以说是无处不在，从手表到超级计算机中都能发现它的身影。Linux 为 Android 提供了硬件抽象层，这使 Android 在未来能够更好地移植到更多的平台上。

从内部来看，Android 的操作系统采用了 Linux 2.6 版的内核，它包括显示驱动、摄像头驱动、蓝牙驱动、USB 驱动、Flash 内存驱动、键盘驱动、WiFi 驱动、Binder 驱动、Audio 驱动以及电源管理驱动。Android 使用 Linux 完成其内存管理、进程管理、网络和其他操作系统的服务工作。Android 手机用户永远也不会看到 Linux，程序也不会直接进行 Linux 调用，但是对于开发人员而言，需要意识到 Linux 在 Android 中的存在。

开发期间需要的某些实用程序要和 Linux 打交道。例如，上一节中提到的"adb shell"命令就可以打开 Linux 的"壳"，从中可以输入要在设备上运行的其他命令，通过这个"壳"来检查 Linux 文件系统、查看活动的进程等。

1.4.2 库

Linux 内核层上面的一层中包含了用 C 或 C++编写的 Android 库文件，这些库文件针对电话使用的特定硬件架构进行了编译，并由手机制造商预先安装到手机中。它们能被 Android 系统中不同的组件使用，它们通过 Android 应用程序框架为开发者提供服务。其中，最重要的库有以下 5 个部分。

1. 外观管理器

Android 使用与 Vista 或 Compiz 类似的组合窗口管理器，只是它将绘制命令传递给屏幕外的位图，然后将该位图与其他位图组合起来，形成用户看到的显示内容。这种方法允许系统实现一些有趣的效果，如透明的窗口和奇特的过渡效果。

2. 2D 和 3D 图形

使用 Android 时，二维和三维元素可结合到一个用户界面中。该库可以使用硬件 3D 加速（如果硬件可用）或者使用高度优化的 3D 软加速。

3. 媒体编解码器

Android 可以播放多种格式的视频内容，并可用各种格式录制和播放音频，同时还支持静态图像文件。常见的编码格式有 aac、avc（H.264）、mp3、amr、mpeg4、jpg 和 png。

4. SQL 数据库

与 Firefox 和苹果的 iPhone 一样，Android 提供了轻量级的 SQLite 数据库引擎。SQLite 功能强大，是一款轻型关系型数据库引擎，使用该引擎可以在应用程序中实现持续存储。

5. 浏览器引擎

为保证快速显示 HTML 的内容，Android 使用了 WebKit 库，用来支持 Android 浏览器和一个可嵌入的 Web 视图。WebKit 是一个开源项目，可以为 Android 内部自带的浏览器所调用。Google 的 Chrome 浏览器、Apple 的 Safari 浏览器、Apple 的 iPhone 和诺基亚的 S60 平台都使用了该引擎。

需要注意的是，这些库不是独立的应用程序，它们只是供高级程序调用。启动 Android

1.5 后，便可使用 NDK（Native Development Toolkit，本机开发工具包）来编写自己的库文件了。如果对这部分特别感兴趣，可以参考 Android 官方提供的 API 或者下载源代码学习。

1.4.3 Android 运行时

在 Linux 内核层上面还有一个 Android 运行时层，该层包括 Dalvik 虚拟机以及 Java 核心库。

Dalvik 是由 Google 的 Dan Bornstein 设计并编写的一款 VM（Virtual Machine，虚拟机）。开发人员编写的代码首先编译为与机器无关的指令，称为字节码，然后由移动设备上的 Dalvik VM 执行这些字节码。虽然不同的字节码格式稍有不同，但 Dalvik 本质上是一个针对低内存耗用而优化的 Java 虚拟机。它允许同时运行多个 VM 实例，并且能够充分利用底层操作系统实现进程隔离。Dalvik 虚拟机专门针对移动设备进行了优化。为 Android 编写的所有代码使用的都是 Java 语言，这些代码都在虚拟机中运行。

Dalvik 与传统的 Java 虚拟机的不同之处体现在两个方面。首先，Dalvik VM 运行 .dex 文件，即编译时会将标准的 .class 和 .jar 文件转换为 .dex 文件。.dex 文件比类文件更加紧凑并且更加高效，这是针对运行 Android 的设备内存有限且通过电池供电的特点所做出的重要改进。其次，Android 附带的 Java 核心库与 Java SE 库和 Java Me 库不尽相同。

1.4.4 应用程序框架

位于核心库和 Android 运行时上面的是应用程序框架层。该层提供了在创建应用程序时需要使用的各种高级构建块。Android 中的应用开发框架设计得非常巧妙，通过这套应用框架，各种组件可以被用户的应用重复利用，各种服务也可以被各种应用重复利用。开发人员只有很好地理解这套框架的工作机制，才能开发出更好的应用程序。

在该框架中，最重要的部分有以下 5 个方面。

1. 活动管理器（Activity Manager）

在 Android 应用中，每一个应用一般是由多个页面组成的，而每个页面的单位就是 Activity。活动管理器的作用在于控制应用程序的生命周期，提供了应用页面退出的机制，同时维护一个公共的"后退栈（backstack）"供用户导航。

2. 内容提供者（Content Provider）

这些对象封装了需要在应用程序之间互访和共享的数据，如联系人信息等。

3. 资源管理器（Resource Manager）

资源是程序中涉及的任何非代码内容，如本地的图片资源、涉及布局的 XML 文件及国际化的字符串等。

4. 位置管理器（Location Manager）

位置管理器用来提供位置服务，所以 Android 手机始终知道它当前所处的位置。

5. 通知管理器（Notification Manager）

通知管理器可以让程序将需要警示的信息显示在状态栏上。例如，当手机用户收到短信、临近预约时间、手机内存临界状态报警、异常入侵等事件发生时，都可以通过友好的方式通知给用户。

1.4.5 应用程序

Android 架构图中最高层是应用程序层。该层就像是 Android 的冰山一角，最终的手机用户只能看到这些应用程序，丝毫不会察觉到在该层下面执行的操作和下面各层提供的服务。这一层的应用程序都是基于 Java 语言而开发，如地图软件、联系人管理、E-mail 连接、浏览器等。许多开发出的程序也都是运行在本层。例如，音乐播放器、通信录、屏幕锁等。

1.5 Android 平台简介

如前所述，Android 平台采用了整合的策略思想，包括底层 Linux 操作系统、中间层的中间件和上层的应用程序。从 2007 年至今，Android 经历了多种版本，由于其固有的平台特征，如今在手机市场中占有率居高不下。

1.5.1 Android 平台的特性

1. 开放性

Android 设计之初首先提倡的就是建立一个标准化、开放式的移动软件平台，所以 Android 操作系统是直接建立在开放源代码的 Linux 操作系统上进行开发。

2009 年，前 Google 中国研究院副院长林斌（现小米公司总裁）在"中国国际信息通信展览会"上曾说过，"Android 平台是全球第一个完整的完全开放的手机平台"。他也特别指出"开放"的两个含义：一方面，Android 手机平台底下的所有核心功能都可以通过标准的接口来调用；另一方面，Android 平台中从最底层的内核到上层的内核等每个环节都是开源的。

正是开放性使得更多的硬件生产商加入到了 Android 开发阵营，也有更多的 Android 开发者投入到了 Android 的应用程序开发中，这些都为 Android 平台带来了大量新的应用。

2. 平等性

所谓平等即意味着 Android 平台对所有的第三方应用软件，以及平台里面自己携带的功能组件一视同仁。例如，如果某手机用户觉得手机主屏不好看，想要进行个性化开发，可以完全拿开源代码把组件进行替换。再例如，某应用程序有问题，自己开发一个也是可以的。如果觉得浏览器不够炫，用户完全可以开发一个自己的浏览器把旧的换掉。

总之，在 Android 操作系统上，所有的应用程序，不管是系统自带的还是由应用程序开发者自己开发的，都可以根据用户的喜好任意替换。

3. 无界性

无界性即数据无间隙整合的特征。在多个应用程序之间，所有的程序都可以方便地进行互相访问一些公共数据，不会受到程序的限制。开发人员可以将自己的程序与其他程序进行交互。例如，通信录的功能本身可以由 Android 手机提供，但是开发人员也可以直接调用通信录的程序代码，并在自己的应用程序上使用。

4. 方便性

Android 使用 Java 作为开发语言，所以对熟悉 Java 的开发人员没有任何难度。在 Android 操作系统中，为用户提供了大量的应用程序组件，如 Google Map、图形用户界面、电话服务等。用户直接在这些组件的基础之上构建自己的开发程序即可。

5. 丰富性

所谓丰富性是指硬件较为丰富。由于平台开放，所以有更多的移动设备厂商根据自己的情况推出了各式各样的 Android 移动设备，虽然在硬件上有一些差异，但这些差异并不会影响数据的同步与软件的兼容，丰富的硬件为不同需求的用户提供了不同的选择。

1.5.2 Android SDK 版本的特点

Android 最早的发布版本开始于 2007 年 11 月的 Android 1.0 beta，迄今为止已经发布了多个更新版本的 Android 操作系统。这些更新版本都在前一个版本的基础上修复了 bug 并且添加了前一个版本所没有的新功能。从 2009 年 4 月开始，Android 操作系统改用甜点来作为版本代号，这些版本按照大写字母的顺序来进行命名：纸杯蛋糕（Cupcake）、甜甜圈（Donut）、闪电泡芙（Éclair）、冻酸奶（Froyo）、姜饼（Gingerbread）、蜂巢（Honeycomb）、冰淇淋三明治（Ice Cream Sandwich）以及果冻豆（Jelly Bean）。此外，Android 操作系统曾经还有两个预发布的内部版本，它们分别是铁臂阿童木（Astro）和发条机器人（Bender）。

Android SDK 各版本的发展进程如图 1-18 所示。

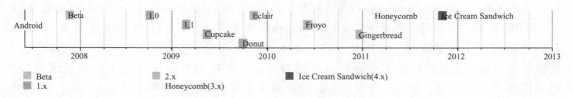

图 1-18 Android SDK 版本进程示意图

1. Android 1. x

Android 1.0 是 Android 操作系统中的第一个正式版本，它于 2008 年 9 月 23 日发布，代号为铁臂阿童木（Astro）。全球第一台 Android 设备 HTC Dream（G1）就是搭载 Android 1.0 操作系统。

2009 年 2 月 2 日，Android 1.1（Bender 发条机器人）发布，该版本只被预装在 T-Mobile G1 上。该版本处理了前一版本遗留的许多应用程序 bug 和系统 bug，改进了 API 接口和添加了新的特性。例如，用户搜索企业和其他服务时，下方会显示出其他用户搜索时对该搜索信息的评价和留言；加强了电话功能，改进了免提功能；支持对邮件附件的保存和预览功能；增加了长按任意界面弹出多选框的功能。

2009 年 4 月 30 日，Android 1.5（Cupcake 纸杯蛋糕）发布，该版本基于 Linux 2.6.27 内核。主要的更新有：能够拍摄/播放影片，并支持上传到 YouTube；支持立体声蓝牙耳机，同时改善自动配对性能；最新的采用 WebKit 技术的浏览器，支持复制/粘贴和页面中搜索；GPS 性能大大提高；提供屏幕虚拟键盘；主屏幕增加音乐播放器和相框 widgets；应用程序自动随着手机旋转；短信、Gmail、日历、浏览器的用户界面大幅改进；相机启动速度加快；来电照片显示等。

2009 年 9 月 15 日，Android 1.6（Donut 甜甜圈）发布，该版本基于 Linux 2.6.29 内核。主要的更新有：重新设计的 Android Market；支持手势；支持 CDMA 网络；文本转语音系统

（Text-to-Speech）；快速搜索框；全新的拍照界面；查看应用程序耗电；支持虚拟私人网络（VPN）；支持更多的屏幕分辨率；支持 OpenCore 2 媒体引擎；新增面向视觉或听觉困难人群的易用性插件等。

2. Android 2.x

2009 年 10 月 26 日，Android 2.0（Éclair 闪电泡芙）版本软件开发工具包发布，该版本基于 Linux 2.6.29 内核。主要的更新有：优化硬件速度；新的浏览器的用户界面和支持 HTML 5；新的联系人名单；更好的白色/黑色背景比率；改进 Google Maps 3.1.2；支持内置相机闪光灯；改进的虚拟键盘；支持支持动态桌面的设计等。

Android 2.1 更新包则于 2010 年 1 月 12 日正式发布。该版本针对 Android 2.0.1 进行了轻微的改进，只针对前一个版本中的部分 API 进行修改变化，并且对存在的已知 bug 进行修复。

2010 年 5 月 20 日，Android 2.2（Froyo 冻酸奶）版本发布，该版本基于 Linux 2.6.32 内核。主要的更新有：支持将软件安装至扩展内存；集成 Adobe Flash 10.1 支持；加强软件即时编译的速度；新增软件启动"快速"至电话和浏览器；USB 分享器和 WiFi 热点功能；支持在浏览器上传档案；更新 Market 中的批量和自动更新；增加对 Microsoft Exchange 的支持；集成 Chrome 的 V8 JavaScript 引擎到浏览器；加强快速搜索小工具；速度和性能优化等。

2010 年 12 月 6 日，Android 2.3（Gingerbread 姜饼）版本发布，该版本基于 Linux 2.6.35 内核。主要更新有：修补 UI；支持更大的屏幕尺寸和分辨率（WXGA 及更高）；系统级复制粘贴；重新设计的多点触摸屏幕键盘；原生支持多个镜头（用于视频通话等）和更多传感器（陀螺仪、气压计等）；电话簿集成 Internet Call 功能；支持近场通信（NFC）；优化游戏开发支持；多媒体音效强化；从 YAFFS 转换到 ext4 文件系统；开放了屏幕截图功能等。

3. Android 3.x

2011 年 2 月 22 日，Android 3.0（蜂巢 Honeycomb）正式发布，该版本基于 Linux 2.6.36 内核，是第一个 Android 平板操作系统。全球第一个使用该版本操作系统的设备是摩托罗拉公司于 2011 年 2 月 24 日发布的 Motorola Xoom 平板电脑。主要更新有：供平板电脑使用的 Android 3.0 界面；支持平板电脑大荧幕、高分辨率；新版 Gmail；3D 加速处理；网页版 Market（Web store）详细分类显示；依个人 Android 分别设定安装应用程序；新的短消息通知功能；加强多任务处理的界面；重新设计适用大屏幕的键盘及复制/粘贴功能；多个标签的浏览器以及私密浏览模式；快速切换各种功能的相机；增强的图库与快速滚动的联络人界面；支持多核心处理器等。

Android 3.1 软件开发包于 2011 年 5 月 10 日正式发布。主要更新有：用户界面的改进；能够连接 USB 配件；扩大最近的应用程序列表；可调整大小的主屏幕小部件；支持外部键盘和鼠标等设备；支持操纵杆和游戏控制器；FLAC 音频播放支持；高性能的 WiFi 锁；支持 HTTP 代理等。

Android 3.2 软件开发包于 2011 年 7 月 15 日正式发布。全球第一台使用该版本操作系统的设备是我国华为公司制造生产的 MediaPad 平板电脑。主要更新有：改进的硬件支持；增强应用程序访问 SD 卡上文件的能力；提供同步功能；增加应用程序兼容性功能；新的显示支持功能；为开发人员提供更多的显示外观在不同的 Android 设备的控制等。

4. Android 4.x

Android 4.0（Ice Cream Sandwich 冰淇淋三明治）于 2011 年 4 月在 Google I/O 大会上首次被宣布，并且于 2011 年 10 月 19 日正式发布 Android 4.0 冰淇淋三明治操作系统和搭载 Android 4.0 的 Galaxy Nexus 智能手机。

Android 4.0.1 的软件开发包于 2011 年 10 月 19 日正式发布。Google 公司的发言人加布·科恩（Gabe Cohen）指出，Android 4.0"理论上"与任何一个当前市面上搭载了 Android 2.3.x 版本的设备兼容。该版本的主要更新内容有：统一了手机和平板电脑使用的系统；应用会自动根据设备选择最佳显示方式；支持在系统中使用虚拟按键；界面以新的标签页形式展示；更方便地在主界面创建文件夹；改进的可视化语音邮件的能力；Gmail 离线搜索；与其他第三方微博、博客类应用程序的无缝连接；实时更新的内容会被展示在主界面上；在锁屏状态下也可以对用户设置的某些应用程序进行操作；面部识别进行锁屏；支持最多同时打开 16 个标签页；自动同步用户手机中的网页书签；可以在桌面版 Chrome 和其他 Android 设备中进行同步；内置流量监控功能；能够随时关闭正在使用的应用程序；提升自带的相机功能；内置图片处理软件；新的图库软件；支持 NFC 功能；新的启动画面；增加支持硬件加速的功能等。

Android 4.1（Jelly Bean 果冻豆）更新包于 2012 年 6 月 28 日在 Google I/O 大会上随搭载 Android 4.1 的 Nexus 7 平板电脑一起发布，主要的更新有：提升用户页面的速度与流畅性；"Google Now"可在 Google 日历内加入活动举办时间、地点；新增脱机语音输入；通知中心显示更多信息；更多的平板优化；Google Play 增加电视片与电影的购买；提升反应速度；强化默认键盘；大幅改变用户界面设计；更多的 Google 云集成；恶意软件的保护措施等。

首款搭载 Android 4.2 的手机 LG Nexus 4 及平板电脑 Nexus 10 于 2012 年 11 月 23 日发售。主要更新有：PhotoSphere 360 度全景拍摄；手势输入键盘；可在屏幕锁定界面直接打开相机功能；Daydream 屏幕保护设备功能；可直接进行操作的状态通知列功能；Miracast 无线显示分享功能；连点三次可放大整个显示页及两指旋转和缩放；增加为盲人用户设计的语音输出及手势模式导航功能；内置时钟新增世界时钟、秒表和定时器；Google Now 新增以 Gmail 登录信息作为数据源；Google Now 新增航班追踪功能、酒店、餐厅预订，与音乐和电影推荐功能等。

2013 年 9 月 3 日，Google 公布 Android 4.3 的后续版本为 4.4（Kit Kat 奇巧巧克力）。2013 年 10 月 31 日，Google 正式发表 Android 4.4（KitKat）版本，以及 Nexus 5。新版本的主要更新功能有：支持语音打开 Google Now；支持全屏模式 Immersive Mode；优化存储器使用；新的电话通信功能；低电耗音乐播放；新的 NFC 付费集成；增加了 ART 模式等。

5. Android 5.x

2014 年 6 月 25 日于 Google I/O 2014 大会上发布 Developer 版（Android L），之后在 2014 年 10 月 15 日正式发布且名称定为 Lollipop，即"棒棒糖"。

在这款全新版本的系统中，采用了全新 Material Design 界面，支持 64 位处理器，全面由 Dalvik 转用 Android RunTime（ART）编译，性能可提升 4 倍；改良了通知界面，新增了 Priority Mode、预载省电及充电预测、自动内容加密、多人设备分享等功能。其中多人设备分享意味着可在其他设备登录自己账号，并获取用户的联系人、日历等 Google 云数据。同时，这个版本的系统还强化了网络及传输连接性，包括 WiFi、蓝牙及 NFC；强化了多媒体功能，例如支持 RAW 格式拍摄；强化了"OK Google"功能。改善了 Google Now 的功能以及对 Android TV 的支持，提供低视力的设置，以协助色弱人士。

根据 OpenSignal 提供的数据显示，在众多版本的系统中，多数 Android 用户依旧选择使用 Ice Cream Sandwich（Android 4.0）版本，约 20.9% 的用户偏向于 KitKat（Android 4.4）版本，只有的 14% 的用户依旧坚持使用 Gingerbread（Android 2.3）或 Froyo（Android 2.2）老版系统，对于"棒棒糖"的尝试者还不多。

1.6 实例 1："你好，Android"

ADT 附带了一个内置的示例程序 Demo，使用这个 Demo 用很短的时间即可快速创建一个简单的"你好，Android！"程序。

1.6.1 创建应用程序

1. 新建 Android 工程

启动"Eclipse"，打开"File"菜单，单击"New"命令，在弹出的子菜单中单击"Android Application Project"命令，紧接着会出现如图 1-19 所示的工程信息窗口，需要输入 Application Name（即应用程序名称）、Project Name（即项目名称）、Package Name（即程序包名称）。其中应用程序名和项目名可以一样，只是应用程序名称在运行时以应用的标题在手机或模拟器中予以显示，而项目名称则是在 Eclipse 的工作空间中作为应用的显示名称。需要注意的是，程序的包名是对应用程序的唯一标识，并且需要至少两个字段构成，字段与字段之间用点号来分隔。

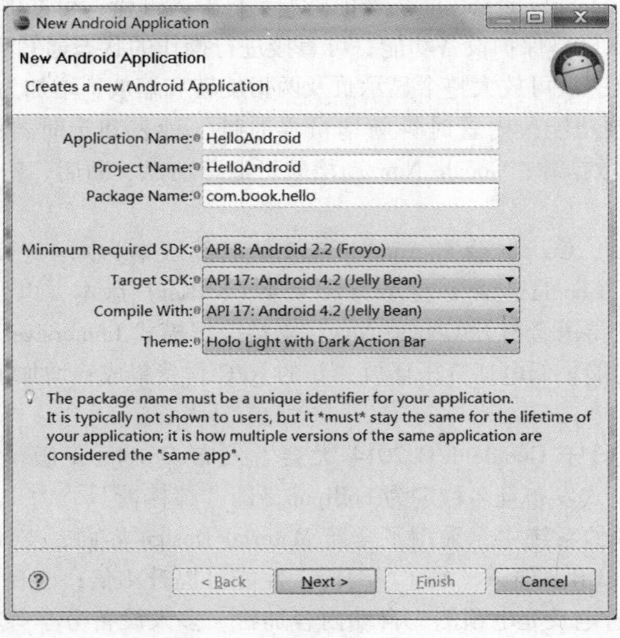

图 1-19 填写 Android 工程信息

可参考如下输入信息：

> Application Name：HelloAndroid
> Project Name：HelloAndroid
> Package Name：com. book. hello

与此同时，需要对一些平台做出选择。Minimum Required SDK 表示在运行时应用程序可以支持的最低目标版本；Target SDK 表示在运行时应用程序能够支持的最高目标版本；Compile With 表示编译应用程序的版本；Theme 表示应用程序使用的基本主题。在做这个项目时，使用默认值即可。

2. 选择项目配置信息

单击"Next"按钮即可进入到如图 1-20 所示的窗口，在这个窗口中需要开发者确定是否使用定制的应用程序图标，是否需要创建 Activity，是否将项目创建在工作空间中等。

图 1-20　选择配置项目

3. 设置项目图标

单击"Next"按钮可以进入到如图 1-21 所示的配置应用程序图标属性的窗口。在这个窗口中需要确定应用程序的图标所用的图形，图标的外形，图标的背景颜色等，这些均可以根据需要进行设置。

4. 选择 Activity 类型

单击"Next"按钮可以进入到如图 1-22 所示的选择 Activity 类型的配置窗口。本例使用默认值 BlankActivity。

5. 为 Activity 和 Layout 命名

单击"Next"按钮可以进入到为 Activity 和 Layout 命名的配置窗口，本例中 Activity 的名称使用默认值"MainActivity"，Layout 的名称使用默认值"activity_main"。输入完毕后，单击"Finish"按钮即可完成项目创建。

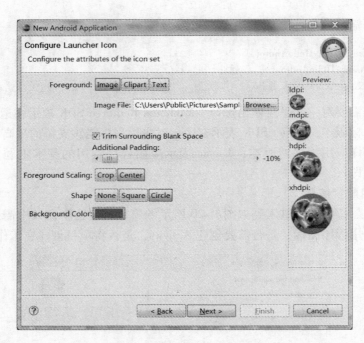

图 1-21　配置应用程序图标

图 1-22　配置 Activity 类型

1.6.2　在模拟器上运行应用程序

1. 运行创建好的应用程序

将项目创建好后，在 Eclipse 中选中该项目，在右键菜单中选择"Run As"命令，然后在弹出的子菜单中选择"Android Application"命令。如果没有创建好 AVD，则运用第 1.2 节的方法创建好 Target 为 SDK 4.2 的模拟器后，再运行，运行效果如图 1-23a 所示，应用程序图标在模拟器中的效果如图 1-23b 中左下角所示。

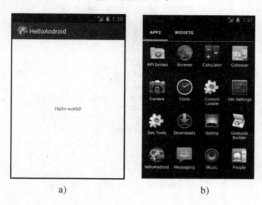

图 1-23　运行效果图

2. 修改应用程序

在运行时，如果想让文字显示"你好，Android！"，则还需要对现有应用程序进行简单修改。在 Eclipse 中选中"HelloAndroid"项目文件夹，打开 res 文件夹中的 values，其中有一个定义了项目中字符串常量的文件 strings.xml，将其中用于显示字符串的值由"Hello world！"更改为"你好，Android！"即可。strings.xml 文件的完整参考代码如下：

```
<?xml version = "1.0" encoding = "utf - 8"?>
<resources>
<string name = "app_name">HelloAndroid</string>
<string name = "hello_world">你好,Android!</string>
<string name = "menu_settings">Settings</string>
</resources>
```

3. 再次运行应用程序

在修改好应用程序之后，保存项目文件。再来运行一次应用程序，即可看到如图 1-24 所示的效果。

需要注意的是，应用程序如果还在模拟器中处于活动状态，没有结束运行，再次运行时在 Console 窗口中会提示"Activity not started, its current task has been brought to the front"的警告信息。此时，只要先结束模拟器中运行的应用程序，再运行就可以了。

图 1-24　修改后的运行效果

1.6.3　Android 应用程序的项目结构

从本例可以看出，在一个项目建立完成后，项目的文件夹结构如图 1-25 所示。

Android 项目包含许多的文件和文件夹，它们都与开发有直接的关系。各个文件夹的作用如表 1-1 所示。

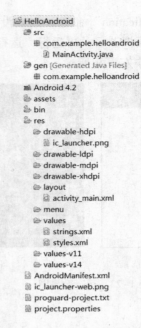

图1-25 Android 项目的文件夹结构

表1-1 Android 项目中文件夹的作用

序号	文件夹名称	用途描述
1	src	存放所有的 Java 源程序
2	gen	存放了 ADT 插件自动生成的代码文件，gen 中的.java 文件是在建立项目时自动生成的，文件均为只读模式，不能更改
3	Android4.2	表示现在使用的 Android SDK 的版本是4.2，如果建立时选择其他版本，则此处显示为其他版本号
4	assets	存放项目中一些较大的资源文件，如图片、音乐、字体等
5	bin	其中存放了项目的可执行文件
6	res	存放项目中所有的资源文件，如图片、网页、文本等
7	res\drawable－hdpi	保存高分辨率的图片资源，可以使用 Resources.getDrawable(id)的方式获得资源，下同
8	res\drawable－ldpi	保存低分辨率的图片资源
9	res\drawable－mdpi	保存中等分辨率的图片资源
10	res\drawable－xhdpi	保存超高分辨率的图片资源，以便使应用适应平板电脑
11	res\layout	存放所有的布局文件，所有的文件均为.xml 格式
12	res\menu	项目中的菜单资源文件必须放在 res\menu 目录中，而且菜单资源文件必须使用 < menu > 标签作为根节点
13	res\values	存放一些资源文件的信息，用于读取文本资源，在本文件夹中有一些约定的文件名称。例如，arrays.xml 是用来定义数组数据的文件名称，colors.xml 是用来定义颜色数据的文件名称，strings.xml 是用来定义字符串数据的文件名称，dimens.xml 是用来定义尺度的文件名称等
14	res\values－v11	代表在 API 11（即 Android 3.0）以上的设备上，用该目录下的 styles.xml 代替 res\values\styles.xml
15	res\values－v14	代表在 API 14（即 Android 4.0）以上的设备上，用该目录下的 styles.xml 代替 res\values\styles.xml

在表 1-1 中，assets 和 res 文件夹都可以存放项目的资源文件，但使用 res 存放则更加方便。这是因为如果将资源保存在 res 文件夹中，ADT 插件会自动帮助用户在 R.java 文件中生成相应的 ID，一旦生成了这些 ID，以后在用户所编写的程序中就可以直接通过 ID 取得各种所需要的资源。而放到 assets 文件夹中的资源文件不会自动生成 ID，访问时不够方便。所以，在开发中，建议使用 res 文件夹保存资源文件。

如表 1-1 所示，在一个 Android 项目中，存在 4 种存放图片资源的文件夹：res\drawable-hdpi、res\drawable-mdpi、res\drawable-ldpi、res\drawable-xhdpi。这是由于支持 Andriod 手机较多，而不同的手机具有不同的屏幕大小，为了图片的显示效果更好，推荐使用这几个文件夹分别存放不同分辨率的图片。

Android 项目中各个文件的作用如表 1-2 所示。

表 1-2 Android 项目中各文件的作用

序号	文件	用途描述
1	MainActivity.java	Activity 程序，是应用程序与用户进行互动的窗口
2	ic_launcher.png	项目中所需要的图片资源文件
3	activity_main.xml	布局文件，用于定义不同显示组件的排列方式
4	strings.xml	用来定义字符串资源的文件名称
5	styles.xml	用来定义样式资源的文件名称
6	AndroidManifest.xml	项目的主要配置文件，配置应用入口，各个组件，以及一些访问权限等。在该文件中，必须声明应用的名称、应用所用到的 Activity 等组件
7	proguard-project.txt 和 project.properties	混码文件，防止他人反编译项目

在 gen 目录下有一个 R.java 文件，它定义了一个 R 类，包含了应用中的各种布局文件、图像、字符串等各种资源与之对应的资源编号（id）。这些资源编号都由系统自动生成，也就是说有一个资源对象，系统就为其在 R 类中生成相应的资源编号，就像一本字典。然而，当 res 文件夹下有任何资源文件有错误时，则无法生成 R 类。所以，当应用程序出现无法生成 R 类的问题时，应马上查看 res 目录下的资源文件是否有误。

1.7 动手实践 1：第 1 个 Android 应用

1.7.1 功能要求

创建应用程序"第 1 个 Android 应用"，要求使用自定义的图标，程序的运行效果以及应用程序在手机应用列表中的效果分别如图 1-26a 和图 1-26b 所示。

1.7.2 操作提示

1. 安装开发环境
在装有 Windows XP 以上操作系统的 PC 上安装 Android SDK 的开发环境。

2. 创建 AVD 并配置
创建 AVD，并启动一个模拟器，运用"Settings"，单击"Language and Input"命令，在

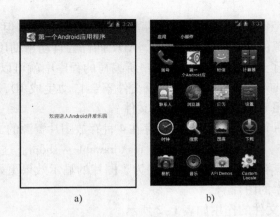

图1-26 习题效果

弹出的对话框中选中"Language"选项,将模拟器的显示语言设置为中文(简体)。

3. 完成应用

本应用可以参考第1.6节内容完成实现。

第 2 章 Android 中的资源

Android 中的资源是在代码中使用的外部文件。这些文件作为应用程序的一部分，被编译到应用程序当中。从上一章的内容可以知道，在创建完 Android 项目后，项目文件夹中有两个存放资源文件的文件夹 res 和 assets。相比之下，assets 中的资源很少用到，而 res 中的资源较为常用。

本章将介绍 Android 中支持的 res 中常用资源文件的定义和使用方法，如颜色资源、字符串资源、布局资源、图片资源、菜单资源等。这些资源文件在项目中有特定的存放位置和格式，详细结构如表 2-1 所示。

表 2-1 Android 资源位置格式表

序 号	目 录 结 构	资 源 格 式
1	res/anim/	XML 动画文件
2	res/drawble/	位图文件
3	res/layout/	XML 布局文件
4	res/values/	arrays.xml：数组文件 colors.xml：颜色文件 dimens.xml：尺寸文件 styles.xml：样式文件 strings.xml：字符串文件
5	res/xml/	任意的 XML 文件
6	res/raw/	直接复制到设备中的原生文件
7	res/menu/	XML 菜单文件

2.1 实例 1：千变万化背景色

2.1.1 功能要求与操作步骤

1. 功能要求

完成如图 2-1a、图 2-1b、图 2-1c 所示的应用程序。要求能够：

① 在应用程序的 res/colors.xml 中定义至少 3 种不同的颜色。
② 运用定义好的颜色使应用程序在每次启动时显示不同的背景色。
③ 运用其中某种颜色设置应用程序中文本的颜色。

2. 操作步骤

（1）创建项目

根据第 1.6.1 节的内容创建 Android 项目，可参考如下输入信息：

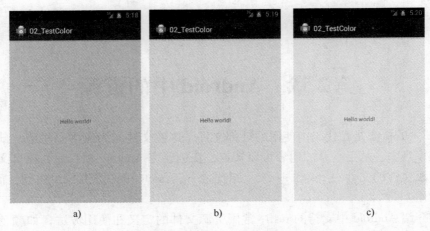

图 2-1 实例 1 运行效果图

```
Application Name：02_TestColor
Project Name：02_TestColor
Package Name：com. book. testcolor
Activity Name：MainActivity
Layout Name：activity_main
```

(2) 创建 colors. xml 文件

单击项目 res/values 目录，在右键菜单中，单击"New"命令，在弹出的子菜单中选中"Other…"命令。然后，在弹出的新建向导中，选择"Android XML Values File"，如图 2-2 所示。

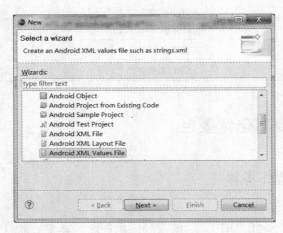

图 2-2 选择 Values 文件向导

单击"Next"按钮，在弹出的新建 Values XML File 窗口中将文件命名为"colors. xml"，如图 2-3 所示。

(3) 定义颜色

在创建好的 colors. xml 文件中，定义 5 种名称分别为"bgcolor1""bgcolor2""bgcolor3""bgcolor4"和"purple"的颜色，其值依次是#fe7（符合#RGB 的格式）、#f5fe（符合

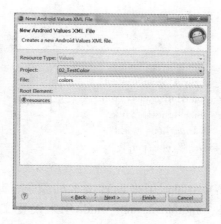

图2-3 为文件命名

#ARGB 的格式)、#f9f0006f (符合#AARRGGBB 的格式)、#9fdd03 (符合#RRGGBB 的格式)和#f0f (符合#RGB 的格式)。参考代码如下：

```
<?xml version = "1.0" encoding = "utf-8"?>
<resources>
<color name = "bgcolor1"> #fe7 </color>
<color name = "bgcolor2"> #f5fe </color>
<color name = "bgcolor3"> #f9f0006f </color>
<color name = "bgcolor4"> #9fdd03 </color>
<color name = "purple"> #f0f </color>
</resources>
```

（4）在 Java 文件中使用颜色

src 目录下的 MainActivity.java 文件中的 onCreate 方法中使用定义好的颜色更改背景色。参考代码如下：

```
protected void onCreate(Bundle savedInstanceState) {
    super.onCreate(savedInstanceState);
    setContentView(R.layout.activity_main);
    //引用颜色资源,将资源 ID 放到数组 bgc 中
    int bgc[] = {R.color.bgcolor1,
                 R.color.bgcolor2,
                 R.color.bgcolor3,
                 R.color.bgcolor4};
    //产生[0,bgc.length)之间的随机数
    Random r = new Random();
    int i = r.nextInt(bgc.length);
    //设置背景色
    getWindow().setBackgroundDrawableResource(bgc[i]);
}
```

（5）在 XML 文件中使用颜色

res/layout 目录下的 activity_main.xml 文件中定义了应用的布局，为了使应用程序中文本的颜色设置为 colors.xml 中定义的 purple 值，需要修改 TextView 的 android:textColor 属性值。

参考代码如下:

```xml
<TextView
    android:layout_width = "wrap_content"
    android:layout_height = "wrap_content"
    android:layout_centerHorizontal = "true"
    android:layout_centerVertical = "true"
    android:text = "@string/hello_world"
    android:textColor = "@color/purple"
/>
```

在上述代码中,android:textColor 用来设置文本的颜色,其值为 colors.xml 中名称为"purple"的颜色。除此之外,android:layout_width 用来设置文本的宽度,android:layout_height 用来设置文本的高度,android:layout_centerHorizontal 用来设置文本水平方向是否居中对齐,android:layout_centerVertical 用来设置文本垂直方向是否居中对齐,android:text 用来设置文本的内容。

2.1.2 颜色(color)资源的定义和使用

1. Android 中的颜色值

Android 中的颜色是用透明度(Alpha)、红色(Red)、绿色(Green)和蓝色(Blue)这 4 个数字值来表示的。其中,透明度衡量了颜色的透明程度,其最小值为 0,最大值为 255。当值为 0 时,表示完全透明,此时红、绿、蓝的值将不起任何作用;当值为 255 时,表示颜色完全不透明;当值为 0~255 之间的值时,表示颜色半透明。

颜色值定义的开始是一个#号,后面是"Alpha – Red – Green – Blue"的格式,合法的表示格式有:#RGB、#ARGB、#RRGGBB、#AARRGGBB。

2. 颜色资源在 XML 文件中的定义

如果可能的话,最好在 XML 资源文件 colors.xml 中定义项目中使用到的所有颜色,这样可以保证以后轻松地更改这些颜色定义。定义颜色资源的基本格式是 <color name = "颜色名称" >#颜色值 </color>。例如在 colors.xml 中可以编写如下代码:

```xml
<?xml version = "1.0" encoding = "utf-8"?>   <!-- XML 版本号和编码方式 -->
<resources>                                  <!-- 资源文件的根元素 -->
    <!-- 定义名称为 background_red 的颜色,值为#f00 -->
    <color name = "background_red" >#f00 </color>
    <!-- 定义名称为 text_green 的颜色,值为#0f0 -->
    <color name = "text_green" >#0f0 </color>
    <!-- 定义名称为 black 的颜色,值为#000 -->
    <color name = "black" >#000 </color>
</resources>                                 <!-- 根元素结束标签 -->
```

在上述代码中,"<!--"和"-->"是 XML 文件的注释符号。

3. 颜色在 Java 中的定义

在 Java 中,可以使用 Color 类的某个静态常量来定义一个 color 对象,例如:

```java
int c = Color.BLUE;  //c 被定义为蓝色
```

如果知道透明度、红色、绿色和蓝色的数值,也可以使用静态工厂方法定义颜色,例如:

  int c = Color. argb(126,255,0,0);//c 被定义为半透明的红色

4. 颜色资源的使用

在 Java 代码中,可以通过"R. color. 颜色名称"的方式使用 colors. xml 中定义的各个颜色值,例如以下代码可以更改应用程序的背景颜色:

  getWindow( ). setBackgroundDrawableResource(R. color. background_red)

在其他 XML 文件(如布局文件)中,可以通过"@color/颜色名称"的方式访问 colors. xml 中定义的各个颜色值,例如:

  android:textColor = "@color/text_green"

2.2 实例 2:屏蔽身份证部分信息

2.2.1 功能要求与操作步骤

实名制越来越普遍,各种证件、账号都与身份证有着密切的联系。注册手机号、银行卡,甚至是网络游戏账号都需要登记身份证。在这些动作中,若有不法分子取得身份证信息,就很可能对用户实施不同程度的权益侵犯。为此,在信息传输和表达的过程中,身份证号中的有些信息是作为掩码加工处理的。本例以此为背景,具体要求和操作步骤如下。

1. 功能要求

完成如图 2-4 所示的应用程序。要求在应用程序中能够屏蔽给定身份证号的出生日期信息,具体能够:

① 在应用程序界面中,显示出"strings. xml"中定义的完整身份证号。

② 屏蔽身份证号中的出生日期信息,并显示在程序界面中。

提示:在 Android 4.2 版本的平台中创建项目时,会在项目的 values 目录下自动生成 strings. xml 文件,所以在本项目中不需要再次创建此文件。

图 2-4 实例 2 运行效果图

2. 操作步骤

(1) 创建项目

创建本项目时,可参考如下输入信息:

  Application Name:02_TestString
  Project Name:02_TestString
  Package Name:com. book. teststring
  Activity Name:MainActivity
  Layout Name:activity_main

（2）在"strings.xml"文件中定义字符串

在 res\values\strings.xml 文件中，定义两个名称分别是"title"和"info"的字符串，其值分别是"来自 string.xml 的字符串："和"110234200112034501"，参考代码如下：

```xml
<?xml version = "1.0" encoding = "utf-8"?>
<resources>
<!--项目原有字符串 -->
<string name = "app_name">02_TestString</string>
<string name = "menu_settings">Settings</string>
<!--自定义的两个字符串 -->
<string name = "title">来自 string.xml 的字符串：</string>
    <string name = "info">110234200112034501</string>
</resources>
```

（3）在 XML 文件中使用字符串

res/layout 目录下的 activity_main.xml 文件中定义了应用的布局。如图 2-4 显示，应用程序需要显示 4 个字符串，所以在 activity_main.xml 文件中需要有 4 个 TextView 组件。

为了使 4 个 TextView 从上到下顺序放置，先将布局方法调整为竖直方向的 LinearLayout（线性布局，后文详解）。具体方法是：双击 res/layout 目录下的 activity_main.xml 文件，在代码视图中，可以看到原有的 RelativeLayout 标签中代码如下：

```xml
<RelativeLayout xmlns:android = "http://schemas.android.com/apk/res/android"
    xmlns:tools = "http://schemas.android.com/tools"
    android:layout_width = "match_parent"
    android:layout_height = "match_parent"
    tools:context = ".MainActivity">
    …
</RelativeLayout>
```

将此"RelativeLayout"标签更改为"LinearLayout"，并将属性参考如下代码修改：

```xml
<LinearLayout xmlns:android = "http://schemas.android.com/apk/res/android"
    xmlns:tools = "http://schemas.android.com/tools"
    android:layout_width = "match_parent"
    android:layout_height = "match_parent"
    tools:context = ".MainActivity"
    android:orientation = "vertical">
    …
</LinearLayout>
```

在添加各种组件时，可以在如图 2-5 所示的 Graphical 视图中拖拽相应组件完成，也可以在代码视图中编写代码实现。

下面首先介绍在 Graphical 视图中添加组件的方法。双击 res/layout 目录下的 activity_main.xml 文件，在 Eclipse 的主窗口中选择图 2-5 下方所示的"Graphical Layout"，选中图 2-5 中箭头所指的 TextView 组件，并将其拖到右侧的布局文件窗口中，即可完成一个 TextView 组件的添加。本例中需要添加 3 个，拖拽 3 次即可。

在添加完成后，可以根据需要在 Properties 面板中设置所加组件的常用属性。如果在开发环境中找不到 Properties 面板，可以打开"Window"菜单，单击"Show View"命令，在弹出的子菜单中选择"Other…"，即可打开如图 2-6 所示的窗口，选择"Properties"选项即可在开发环境下方看到如图 2-7 所示的属性面板。

提示：其他面板的打开方式与此相似。

图 2-5 Graphical 视图中添加 TextView 组件

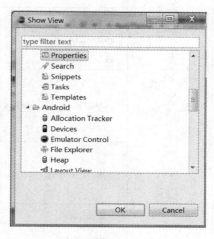
图 2-6 打开各种面板的窗口

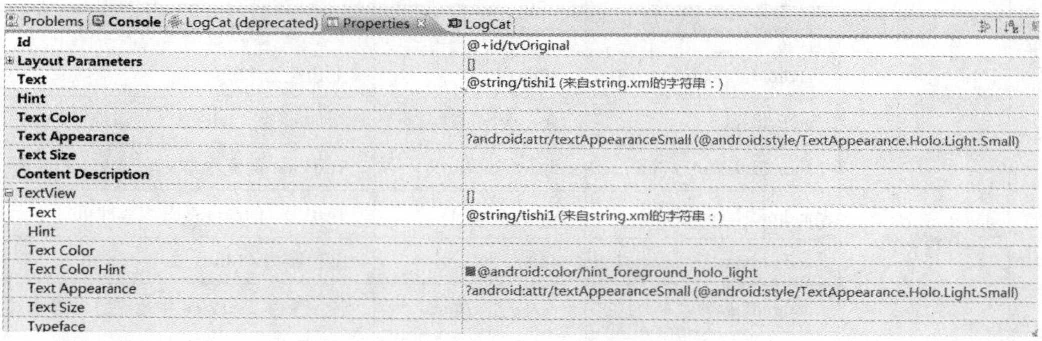

图 2-7 Properties 面板

从上到下依次选择 4 个 TextView，分别在 Properties 面板中对其常用属性进行设置。属性列表可参考表 2-2。

表 2-2 TextView 组件的属性列表

序号	组件类型	属性名称	属性值	说明
1	TextView	android:id	@+id/title	@+id 表示增加 id 为"title"的组件 注意：@id 表示访问 id，@+id 表示增加组件 id
		android:text	@string/title	TextView 的显示文本是 strings.xml 中定义的名称为 title 的字符串
		android:layout_width	match_parent	TextView 的宽度与父容器大小相匹配
		android:layout_height	wrap_content	TextView 的高度适合内容即可

(续)

序号	组件类型	属性名称	属性值	说明
2	TextView	android:id	@+id/tvInfo1	增加 id 为 "tvInfo1" 的组件
		android:text	@string/info	TextView 的显示文本是 strings.xml 中定义的名称为 info 的字符串
		android:layout_width	match_parent	TextView 的宽度与父容器大小相匹配
		android:layout_height	wrap_content	TextView 的高度适合内容即可
		android:gravity	center_horizontal	使文字在 TextView 中水平方向居中对齐
		android:layout_marginTop	16dp	上边距为 16dp。dp 即 dip（device independent pixels），表示设备独立像素。在表示长度、高度属性时可以使用 dp，它与密度无关，与比例有关
		android:textSize	20sp	文字大小为 20sp。sp（scaled pixels）即放大像素，主要用于字体显示。它不仅与密度无关，还与 scale 也无关
2	TextView	android:id	@+id/title_change	增加 id 为 "title_change" 的组件
		android:text	经修改后的字符串：	TextView 显示文本 "经修改后的字符串："
		android:layout_width	wrap_content	宽度适合文字即可
		android:layout_height	wrap_content	TextView 的高度适合文字即可
		android:layout_marginTop	32dp	上边距为 32dp
3	TextView	android:id	@+id/tvInfo2	增加 id 为 "tvInfo2" 的组件
		android:layout_width	match_parent	TextView 的宽度与父容器大小相匹配
		android:layout_height	wrap_content	TextView 的高度适合内容即可
		android:layout_marginTop	48dp	上边距为 48dp
		android:gravity	center_horizontal	使文字在 TextView 中水平方向居中对齐
		android:textSize	20sp	文字大小为 20sp

在代码视图下，可以看到 4 个 TextView 的属性设置代码如下：

```xml
<TextView
    android:id = "@+id/title"
    android:layout_width = "match_parent"
    android:layout_height = "wrap_content"
    android:text = "@string/title" />
<TextView
    android:id = "@+id/tvInfo1"
    android:layout_width = "match_parent"
    android:layout_height = "wrap_content"
    android:layout_marginTop = "16dp"
    android:gravity = "center_horizontal"
    android:text = "@string/info"
```

```
            android:textSize = "20sp"/>
        < TextView
            android:id = "@ + id/title_change"
            android:layout_width = "wrap_content"
            android:layout_height = "wrap_content"
            android:layout_marginTop = "32dp"
            android:text = "经修改后的字符串:"/>
        < TextView
            android:id = "@ + id/tvInfo2"
            android:layout_width = "match_parent "
            android:layout_height = "wrap_content"
            android:layout_marginTop = "48dp"
            android:gravity = "center_horizontal"
            android:textSize = "20sp"
            />
```

(4) 在 Java 文件中使用字符串

src 目录下的 MainActivity.java 文件首先取得了 strings.xml 中定义的字符串 info，然后使用字符串处理函数对 info 进行了加工，屏蔽了与出生日期相关的字符。

为了访问 TextView 组件和定义好的字符串，需要在 MainActivity.java 中首先定义一些相关变量，参考代码如下：

```
    public class MainActivity extends Activity {
        //定义 TextView 对象,将处理后的字符串最终在其中显示
        TextView tv;
        //定义字符串变量,s 用来存放原始字符串,r 用来存放替代字符
          String s,r = " ";
        //定义字符替代起点和终点的 int 型变量
        int start = 6, end = 14;
        protected void onCreate( Bundle savedInstanceState) {
            ...
        }
    }
```

然后，在 onCreate 方法中对字符串进行替换处理。参考代码如下：

```
    protected void onCreate( Bundle savedInstanceState) {
        super.onCreate(savedInstanceState);
        setContentView(R.layout.activity_main);

            //找出 id 为 R.id.tvInfo2 的 TextView,存放改变后的文本
            tv = (TextView) findViewById(R.id.tvInfo2);

            //获取字符串资源 R.string.info 的值,位于 res/values/strings.xml 中
            s = getResources().getString(R.string.info);

            //生成替代字符串 r
            for( int i = 0;i < end - start;i ++ )
```

```
                r = r + " * ";

            //用 r 替换原字符串中的部分内容
            s = s.replace(s.substring(start,end),r);

            //为文本框设置文本内容
            tv.setText(s);
    }
```

在上述代码中使用了 String 类提供的获取子字符串方法 substring(int startIndex, int endIndex)和字符串替换方法 replace(CharSequence target,CharSequence replacement)。

需要说明的是,通过 String 类的 substring()方法可以根据字符串的下标(从 0 开始)对字符串进行截取。Substring()方法被两种不同的方法重载,此处使用的 substring(int startIndex, int endIndex) 方法返回的是从字符串 startIndex 下标位置开始截取至 endIndex 下标位置结束的子字符串。

通过 String 类的字符串替换方法 replace(CharSequence target,CharSequence replacement)可以将 target 所表示的字符串替换成 replacement 字符串的值。如果 target 字符串没有出现字符串序列中,则将原字符串返回。

2.2.2 字符串(string)资源的定义与使用

1. 字符串资源在 XML 中的定义

在一个 Android 项目中,开发人员会用到大量的字符串作为提示信息。为了以后维护应用程序的方便,这些字符串都可以作为字符串资源声明在配置文件 strings.xml 中。定义字符串资源的基本格式是:

```
<string name = "字符串名称">字符串值</string>
```

例如,在 strings.xml 中可以编写如下代码:

```
<?xml version = "1.0" encoding = "utf-8"?>      <!-- xml 版本号和编码方式 -->
<resources>
    <!-- 定义名称为"string1"的字符串,值为"测试字符串 1" -->
    <string name = "string1">测试字符串 1</string>
    <!-- 定义名称为"string2"的字符串,值为"测试字符串 2" -->
    <string name = "string2">测试字符串 2</string>
</resources>
```

2. 字符串在 Java 中的定义

在 Java 中,字符串被作为对象来管理,因此可以像创建其他类的对象一样,来创建字符串对象。创建字符串对象时要使用 String 类的构造方法。例如:

```
String s = new String("good");//字符串 s 的值为"good"
```

除此之外,还可以通过字符串常量的引用赋值给一个字符串变量。例如:

```
String s = "Welcome to China";//字符串 s 的值为"Welcome to China"
```

3. 字符串的使用

在 Java 代码中，可以通过 "R.string.字符串名称" 的方式访问 strings.xml 中定义的各个字符串，在需要取得字符串值时，必须通过 "getResources().getString(R.string.字符串名称)" 方法来获取。例如，以下代码即可取得 strings.xml 中定义的 app_name 的字符串值，并将此值赋予字符串变量 s：

> String s = getResources().getString(R.string.app_name)；

在其他 XML 文件中，可以通过 "@string/字符串名称" 的方式访问 strings.xml 中定义的各个字符串，例如以下代码可以将某组件的显示标题设置为 strings.xml 中定义的 title 的字符串值：

> android：text = "@string/title"

2.3 实例3：渐现"四书五经"

2.3.1 功能要求与操作步骤

1. 功能要求

完成如图 2-8a、b 所示的应用程序启动动画。要求应用程序能够：

① 逐渐由浅入深显示"四书五经"背景图片。
② 将"联系我们"文本和"QQ：12345678"文本在界面的右下角显示。
③ 运行 5 s 后自动退出。

图 2-8 线性布局的应用示例图

2. 操作步骤

（1）创建项目

参考如下输入信息创建 Android 项目：

> Application Name：03_TestLinearLayout
> Project Name：03_TestLinearLayout

> Package Name：com. book. testlinearlayout
> Activity Name：MainActivity
> Layout Name：activity_main

（2）修改布局管理器

在 Android 4.2 及以上版本中，默认的布局方式是 RelativeLayout，故在本例中，需要首先将 activity_main.xml 布局中的"RelativeLayout"替换为"LinearLayout"，然后可以参考表 2-3 设置其属性。

表 2-3 实例 3 中 LinearLayout 的属性列表

序号	属性名称	属性值	说明
1	android：id	@+id/bg	增加 id 为"bg"的组件
2	android：layout_width	match_parent	宽度与屏幕宽度相同
3	android：layout_height	match_parent	高度与屏幕高度相同
4	android：orientation	vertical	竖直方向排列组件
5	android：background	@drawable/sswj	背景图片来自于 res/drawable 中的 sswj.jpg。注意：在使用前需要将此图片先复制到 res/drawable 目录下
6	android：gravity	right|bottom	使组件置于布局的右下方

本例中使用到了图片资源，在此作一说明。Android 中主要支持的图片格式有 png、jpg 和 gif，在使用前需要将图片先置于 res/drawable 中。Java 代码中使用图片时可以通过 Resources.getDrawable(int id) 方法获得，参数为图片资源 id，通过"R.drawable.图片资源名称"来引用。在 XML 文件中使用时，通过"@drawable/图片资源名称"的直接引用即可。

参考代码如下：

```
<LinearLayout xmlns:android="http://schemas.android.com/apk/res/android"
    android:id="@+id/bg"
    android:layout_width="match_parent"
    android:layout_height="match_parent"
    android:orientation="vertical"
    android:background="@drawable/sswj"
    android:gravity="right|bottom"
    >
    <!--定义组件-->
    ...
</LinearLayout>
```

（3）添加组件

在布局管理器的 Graphical 视图中，可以运用拖拽的方式添加两个 TextView 组件；也可以在代码视图下定义这两个组件，其各自属性可参考表 2-4 设置。

表 2-4　实例 3 中两个 TextView 的属性列表

序号	组件类型	属性名称	属性值	说明
1	TextView	android:id	@+id/tv_tip	增加 id 为"tv_tip"的组件
		android:layout_width	wrap_content	宽度适合内容即可
		android:layout_height	wrap_content	高度适合内容即可
		android:text	联系我们	显示文本为"联系我们"
		android:textSize	20sp	文本大小为 20sp
		android:textColor	#000	文本颜色为黑色
2	TextView	android:id	@+id/tv_qq	增加 id 为"tv_qq"的组件
		android:layout_width	wrap_content	宽度适合内容即可
		android:layout_height	wrap_content	高度适合内容即可
		android:text	QQ:12345678	显示文本为"QQ:12345678"

参考代码如下：

```xml
<!-- 定义组件 -->
<TextView
    android:id = "@+id/tv_tip"
    android:layout_width = "wrap_content"
    android:layout_height = "wrap_content"
    android:text = "联系我们:"
    android:textSize = "20sp"
    android:textColor = "#000"/>
<TextView
    android:id = "@+id/tv_qq"
    android:layout_width = "wrap_content"
    android:layout_height = "wrap_content"
    android:text = "QQ:12345678"/>
```

（4）实现 MainActivity 中的功能

渐隐渐现动画可以借助 android.view.animation 中 AlphaAnimation 类的 alpha 属性来实现。例如，在构造 AlphaAnimation 对象时，设置 alpha 值从 0（表示全透明）到 1（不透明）即可让添加动画的对象渐现；反之，则渐隐。在任何一个组件添加动画效果时，都可调用 View 类的 setAnimation(Animation animation)方法。在动画结束后可以通过调用 Activity 的 finish()方法结束 Activity，也可以通过 System.exit(0)退出应用程序。

参考代码如下：

```java
public class MainActivity extends Activity {
    @Override
    protected void onCreate(Bundle savedInstanceState) {
        super.onCreate(savedInstanceState);
        setContentView(R.layout.activity_main);
        //设置透明度区间,实现渐现效果
        AlphaAnimation animation = new AlphaAnimation(0.0f,1.0f);
        //设置动画时间
```

```
                    animation.setDuration(5000);
                    //为动画添加监听
                    animation.setAnimationListener(new AnimationListener(){
                        public void onAnimationEnd(Animation animation){
                            //动画结束后退出本应用
                            System.exit(0);
                        }
                        @Override
                        public void onAnimationRepeat(Animation animation){
                            // TODO Auto-generated method stub
                        }
                        @Override
                        public void onAnimationStart(Animation animation){
                            // TODO Auto-generated method stub
                        }
                    });
                    //取得布局中定义的LinearLayout
                    LinearLayout linearlayout = (LinearLayout) this.findViewById(R.id.bg);
                    //为布局添加动画
                    linearlayout.setAnimation(animation);
                }
            }
```

Android 4.0 以上的版本中常用的布局管理器有 5 种，分别是线性布局、帧布局、表格布局、相对布局和网格布局。在 Android 2.3.3 版本之前，还存在绝对布局（AbsoluteLayout），它使用 X、Y 轴坐标的形式排列组件，由于在 Android 2.3.3 之后不再支持此布局管理器，故本书中不再详细介绍。

不论定义上述哪种布局文件，都需要遵循如下定义格式：

```
<布局类
    xmlns:android = "http://schemas.android.com/apk/res/android"
    布局属性的设置>
        …
</布局类>
```

例如，下述代码定义了一个与屏幕同宽、同高且竖直方向排列组件的线性布局。

```
<LinearLayout
    xmlns:android = "http://schemas.android.com/apk/res/android"
    android:layout_width = "match_parent"
    android:layout_height = "match_parent"
    android:orientation = "vertical" >
    …
</LinearLayout>
```

后文依次介绍线性布局、帧布局、表格布局、相对布局和网格布局的使用方式和常用属性的设置方法。

2.3.2 线性布局（LinearLayout）的定义与使用

线性布局是 Android 中最常用的布局之一，用 LinearLayout 类来表示。它将自己包含的子元素按照水平或竖直方向依次排列，每个子元素都位于前一个元素之后。

如果需要将方向设置为竖直方向，可以将 orientation 的值设置为"vertical"。此时，布局是一个 N 行单列的结构，不论元素的宽度为多少，每行只会有一个子元素。如需将方向设置为水平，可以将 orientation 值设置为"horizontal"。此时，布局是一个单行 N 列的结构。

需要注意的是，无论水平方向还是竖直方向，如果放置的子元素的长度超出了屏幕的宽度，那么超出的子元素将不可见。这是因为线性布局不会自动换行。

LinearLayout 的常用 XML 属性和相关说明见表 2-5 所示。

表 2-5　LinearLayout 的常用 XML 属性和相关说明

序号	XML 属性	说　　明	编码举例
1	android：orientation	设置布局内子元素的排列方向 可选值：vertical（竖直方向）、horizontal（水平方向） orientation 的默认值为 horizontal	android：orientation = "vertical"
2	android：gravity	设置布局内子元素的对齐方式 可选值：top、bottom、left、right、center_vertical、center_horizontal、center、fill、clip_vertical 和 clip_horizontal 可以同时制定多种堆砌方式的组合	android：gravity = "left | center_vertical" 表示布局内子元素在水平方向左对齐，同时竖直方向居中对齐
3	android：layout_width	设置布局的宽度 可选值：match_parent、fill_parent、wrap_content	android：layout_width = "match_parent" 设置布局宽度与父容器宽度相同
4	android：layout_height	设置布局的高度 可选值：match_parent、fill_parent、wrap_content	android：layout_height = "wrap_content" 设置布局高度能够包含内容即可

另外，在 LinearLayout 中子元素的属性 android：layout_weight 生效，它用于描述该子元素在剩余空间中占有的大小比例。使用 weight 属性进行布局，在界面上可以很好地兼容不同分辨率的手机。以水平方向的线性布局为例，假设需要设置子元素的大小比例，那么在使用这个属性时，如果将元素的 android：layout_width 属性设置为"0dp"（官方推荐），则子元素所占空间就按照所设置的 weight 比例进行显示；但是如果子元素的 android：layout_width 属性设置为"wrap_content"，则子元素所占空间表现为 weight 属性的数值越大，所占比例越高。

例如，需要设计如图 2-9 所示的两个不等长的文本框，并且第 1 个文本框占据 2/3，第 2 个文本框将占据 1/3。那么，按照官方意见，在布局时首先将两个文本框的 android：layout_width 属性设置为"0dp"，然后将其 android：layout_weight 属性分别设置为 2 和 1 即可。

图 2-9　weight 属性设置示意图

参考代码如下：

```xml
<LinearLayout xmlns:android="http://schemas.android.com/apk/res/android"
    android:layout_width="fill_parent"
    android:layout_height="fill_parent"
    android:orientation="horizontal" >
    <TextView
        android:id="@+id/textView1"
        android:layout_width="0dp"
        android:layout_height="wrap_content"
        android:text="TextView1"
        android:layout_weight="2"/>
    <TextView
        android:id="@+id/textView2"
        android:layout_width="0dp"
        android:layout_height="wrap_content"
        android:text="TextView2"
        android:layout_weight="1"/>
</LinearLayout>
```

相类似地，如果设置子元素在竖直方向上的剩余比例，则需要将 android:layout_height 属性值进行设置，然后再设置其 weight 属性值即可，原理同上。

2.4 实例4：初读"大学"

2.4.1 功能要求与操作步骤

1. 功能要求

完成如图 2-10 所示的应用程序。要求能够：
① 在界面顶端居中显示前景图片。
② 背景颜色设置为 "#a7926b"。
③ 显示"大学"的内容片段。

2. 操作步骤

（1）创建项目

参考如下输入信息创建 Android 项目：

Application Name：04_TestFrameLayout
Project Name：04_TestFrameLayout
Package Name：com.book.testframelayout
Activity Name：MainActivity
Layout Name：activity_main

图 2-10　帧布局的应用示例图

（2）修改布局管理器

首先将 activity_main.xml 布局中的 "RelativeLayout" 替换为 "FrameLayout"，然后参考表 2-6 设置其属性。

表 2-6　实例 4 中 FrameLayout 的属性列表

序号	属性名称	属性值	说明
1	android:layout_width	match_parent	宽度与屏幕宽度相同
2	android:layout_height	match_parent	高度与屏幕高度相同
3	android:foreground	@drawable/dx	前景图片设置为 res/drawable 中的 dx.png
4	android:foregroundGravity	clip_horizontal|top	clip_horizontal 表示水平方向剪切，即当对象边缘超出容器的时候，将左右边缘超出的部分剪切掉；top 表示将前景图片置于布局顶端
5	android:background	#a7926b	设置背景颜色为#a7926b，使之与前景图片的背景色一致

参考代码如下：

```
< FrameLayout xmlns:android = "http://schemas.android.com/apk/res/android"
    android:layout_width = "match_parent"
    android:layout_height = "match_parent"
    android:foreground = "@drawable/dx"
    android:foregroundGravity = "clip_horizontal|top"
    android:background = "#a7926b"
>
    <!-- 定义组件 -->
    …
</FrameLayout >
```

（3）定义字符串资源

在 res\values\strings.xml 文件中，定义名称为 "dx" 的字符串，值为大学中的内容片段 "大学之道，在明明德。…"。为了能够分段显示，在每段开始时产生缩进效果，在定义字符串时，可以使用转义字符 "\n" 和 "\t"。参考代码如下所示：

```
<?xml version = "1.0" encoding = "utf-8"?>
< resources >
    <!-- 项目原有字符串 -->
    < string name = "app_name" >04_TestFrameLayout </string >
    < string name = "action_settings" >Settings </string >
    <!-- 新增字符串 -->
    < string name = "content" >
        \t 大学之道,在明明德,在亲民,在止于至善。
        \n\t 知止而后有定…
    </string >
</resources >
```

（4）添加组件

在布局管理器的 Graphical 视图中，可以通过拖拽在 activity_main.xml 中添加一个 TextView 组件；也可以在代码视图下定义这个组件，其属性设置参考表 2-7。

表 2-7 实例 4 中 TextView 的属性列表

序号	组件类型	属性名称	属性值	说明
1	TextView	android:id	@+id/tv_content	增加 id 为"tv_content"的组件
		android:layout_width	match_parent	宽度与屏幕宽度相同
		android:layout_height	wrap_content	高度适合内容即可
		android:text	@string/content	显示 string 中定义的 content 文本
		android:textSize	16sp	文本大小为 16sp
		android:layout_gravity	top	将文本框置于布局顶端
		android:layout_marginTop	90dp	文本框与布局顶部之间的边距是 90dp 注意：在 FrameLayout 中，只有在设置了 android:layout_gravity 属性之后，与边距有关的属性才起作用

参考代码如下：

```xml
<!--定义组件 -->
<TextView
    android:id = "@+id/tv_content"
    android:layout_width = "match_parent"
    android:layout_height = "wrap_content"
    android:text = "@string/content"
    android:textSize = "16sp"
    android:layout_gravity = "top"
    android:layout_marginTop = "90dp"/>
```

2.4.2 帧布局（FrameLayout）的定义与使用

帧布局是五大布局中最简单的一个布局，用 FrameLayout 类来表示，在这种布局中，整个屏幕被当成一块空白备用区域，所有的组件都将放在屏幕的左上角。由于无法为这些元素指定一个确切的位置，所以从视觉上看，后面的子元素直接覆盖在前面的子元素之上，将前面的子元素部分和全部遮挡。

除 layout_width 和 layout_height 与 LinearLayout 的属性意义相同之外，FrameLayout 还有其自己常用的 XML 属性，具体说明见表 2-8 所示。

表 2-8 FrameLayout 的常用属性举例说明

序号	XML 属性	说明		编码举例
1	android:foreground	设置帧布局的前景图像		android:foreground = "@drawable/ic_launcher"，表示帧布局的前景图像设置为 drawable 目录下的 ic_launcher 图片
		可选值：图片文件资源		
2	android:foregroundGravity	设置前景图像在帧布局中的位置		android:foregroundGravity = "center_vertical|left"，表示布局中的图片水平方向左对齐，竖直方向居中对齐
		可选值：top、bottom、left、right、center_vertical、center_horizontal、center、fill、clip_vertical 和 clip_horizontal		
		可以同时制定多种堆砌方式的组合		

2.5 实例 5：办公电话一览

2.5.1 功能要求与操作步骤

1. 功能要求

完成如图 2-11 所示"办公电话一览表"的应用程序。要求能够：

① 顶端显示字号为 20sp 的文本"办公电话一览表"。
② 用 3 列若干行的表格完成布局。
③ 表头背景色为黑色，文本为白色，字号大小是 20sp，在单元格中居中对齐。
④ 表中各列具体信息所在单元格加黑色边框，文本为黑色，在单元格中左对齐。

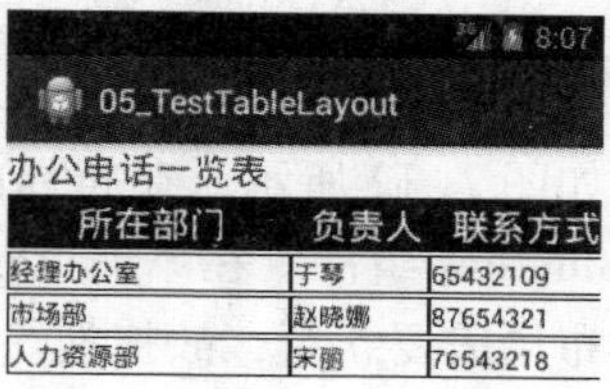

2. 操作步骤

（1）创建项目

参考如下输入信息创建 Android 项目：

```
Application Name：05_TestTableLayout
Project Name：05_TestTableLayout
Package Name：com.book.testtablelayout
Activity Name：MainActivity
Layout Name：activity_main
```

图 2-11 表格布局的应用示例图

（2）修改布局管理器

首先将 activity_main.xml 布局中的"RelativeLayout"更改为"TableLayout"，然后参考表 2-9 设置其属性。

表 2-9 实例 5 中 TableLayout 的属性列表

序号	属性名称	属性值	说明
1	android:layout_width	match_parent	宽度与屏幕宽度相同
2	android:layout_height	match_parent	高度与屏幕高度相同
3	android:stretchColumns	*	表中各列均允许被拉伸，使用这个属性时需要表格中至少包含一列
4	android:background	#fff	设置背景颜色为#fff（白色）

参考代码如下：

```
<TableLayout xmlns:android = "http://schemas.android.com/apk/res/android"
    android:layout_width = "match_parent"
    android:layout_height = "match_parent"
    android:stretchColumns = " * "
    android:background = " #fff"
```

```
    >
        <!--添加行,在行中定义组件-->
        ...
</TableLayout>
```

(3) 添加行

如图 2-12 所示,在布局的 Graphical 视图模式中可以看到"Layouts"选项,其中包含了 TableRow 组件。在添加行时,只需要将此组件拖拽到布局文件中即可。从图 2-11 可看出本例布局需要 5 行,故拖拽 5 次即可。

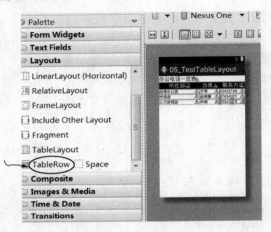

图 2-12 在"Graphical"视图下添加行

为了实现如图 2-11 所示的边框效果,在本例中第 1 行不需要设置任何属性,其余各行也仅需要设置背景颜色和边距。各属性可参考表 2-10 设置。

表 2-10 实例 5 中除第 1 行外的 TableRow 属性列表

序号	属性名称	属性值	说明
1	android:background	#000	设置背景颜色为#000(黑色),与表格的背景色形成反差
2	android:layout_margin	0.5 dip	设置边距为 0.5 dip,实现边框效果的关键

参考代码如下:

```
<!--第1行-->
<TableRow>
    <!--添加组件-->
    ...
</TableRow>
<!--第 2~5 行均如此设置 TableRow 属性-->
<TableRow
    android:background = "#000"
    android:layout_margin = "0.5dip" >
    <!--添加组件-->
    ...
</TableRow>
```

（4）添加组件

在布局管理器的 Graphical 视图中，可以通过拖拽在每个 TableRow 中添加一定数量的 TextView 组件；也可以在代码视图下定义这些组件。其属性设置参考表 2-11。

表 2-11　实例 5 中 TextView 的属性列表

序号	组件类型	属性名称	属性值	说明
1	首行 TextView	android:text	办公电话一览表	显示文本为"办公电话一览表"
		android:textSize	20 sp	设置字号为 20 sp
2	表头 TextView	android:text	所在部门（或负责人、联系方式）	显示文本为"所在部门"或负责人、联系方式
		android:layout_gravity	center	将文本在单元格内居中
		android:background	#000	背景颜色为黑色
		android:layout_margin	0.5 dip	设置边距为 0.5 dip，实现边框效果的关键
		android:textColor	#fff	文本颜色为白色
		android:textSize	20 sp	字号大小 20 sp
3	内容 TextView	android:text	经理办公室（或其他）	显示文本为"经理办公室"或其他
		android:background	#fff	背景颜色为白色
		android:layout_margin	0.5 dip	设置边距为 0.5 dip，实现边框效果的关键
		android:textColor	#000	文本颜色为黑色

表格布局中第 1 行和第 2 行的参考代码如下：

```
<!--第1行 -->
<TableRow>
<!--添加组件 -->
<TextView
        android:text = "办公电话一览表"
        android:textSize = "20sp"/>
</TableRow>
<!--第2行 -->
  <TableRow
     android:background = "#000"
     android:layout_margin = "0.5dip" >
<!--添加组件 -->
      <TextView
android:text = "所在部门"
android:layout_gravity = "center"
android:background = "#000"
android:layout_margin = "0.5dip"
android:textColor = "#fff"
android:textSize = "20sp"/>
      <TextView
android:text = "负责人"
```

```
            android:layout_gravity = "center"
            android:background = "#000"
            android:layout_margin = "0.5dip"
            android:textColor = "#fff"
            android:textSize = "20sp"/>
         <TextView
            android:text = "联系方式"
            android:layout_gravity = "center"
            android:background = "#000"
            android:layout_margin = "0.5dip"
            android:textColor = "#fff"
            android:textSize = "20sp"/>
      </TableRow>
```

其他行可参考以上代码实现，只是需要更改各自的 text 属性值。

2.5.2 表格布局（TableLayout）的定义与使用

表格布局是采用表格的形式对组件的布局进行管理，用 TableLayout 类来表示。TableLayout 不是通过声明包含多少行、列来定义表格，而是通过添加 TableRow 控制表格中的行数，通过添加组件的个数控制表格的列数。也就是说，每当在 TableLayout 中添加一个 TableRow，就会在表格中添加一行；每当为 TableRow 中添加一个组件，就会在该行添加一列。

需要注意的是，TableRow 是 LinearLayout 的子类，因此可以不断地在 TableRow 中添加其他组件，每添加一个子组件该表格就增加一列。只是 TableRow 的 android:orientation 值恒为 "horizontal"，android:layout_width 值恒为 "match_parent"，android:layout_height 值恒为 "wrap_content"。所以 TableRow 中的组件都是横向排列，并且宽高一致。这样的设计使得每个 TableRow 里的组件都相当于表格中的单元格，单元格可以为空，但是不能跨列。

TableLayout 的常用 XML 属性和相关说明见表 2-12 所示。

表 2-12 TableLayout 的常用 XML 属性和相关说明

序号	XML 属性	说　明	编码举例
1	android:collapseColumns	设置需要被隐藏的列的列序号，多个列序号之间用逗号隔开	android:collapseColumns = "1"，表示隐藏第 2 列
2	android:shrinkColumns	设置允许被收缩的列的列序号，多个列序号之间用逗号隔开	android:shrinkColumns = "1"，表示第 2 列收缩
3	android:stretchColumns	设置允许被拉伸的列的列序号，多个列序号之间用逗号隔开	android:stretchColumns = "2"，表示第 3 列拉伸
4	注意：表格布局中的各列序号从 0 开始		

2.6 实例 6：梅花效果首界面

2.6.1 功能要求与操作步骤

1. 功能要求

完成如图 2-13 所示的应用程序。要求能够：

① 在界面顶端居中显示"梅花布局效果"文本，文本颜色为#ff0080，大小为20sp。
② 屏幕中心显示"进入"文本及其背景图，周边围绕显示其余4个文本："帮助""联络""设置""退出"及其背景图。

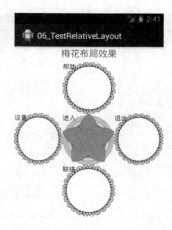

图 2-13　相对布局的应用示例图

2. 操作步骤

（1）创建项目

参考如下输入信息创建 Android 项目：

> Application Name：06_TestRelativeLayout
> Project Name：06_TestRelativeLayout
> Package Name：com.book.testrelativelayout
> Activity Name：MainActivity
> Layout Name：activity_main

（2）设置布局管理器属性

参考表 2-13 设置其属性。

表 2-13　实例 6 中 RelativeLayout 的属性列表

序号	属性名称	属性值	说明
1	android：layout_width	match_parent	宽度与屏幕宽度相同
2	android：layout_height	match_parent	高度与屏幕高度相同

参考代码如下：

```
< RelativeLayout xmlns:android = "http://schemas.android.com/apk/res/android"
    android:layout_width = "match_parent"
    android:layout_height = "match_parent"
    >
    <!--定义组件-->
    ...
</ RelativeLayout >
```

（3）添加组件

从图 2-12 可看出在此布局中，至少需要 6 个 TextView，其宽度和高度属性均可设置为"wrap_content"，其余属性可参考表 2-14 设置。

表 2-14　实例 6 中各 TextView 的属性列表

序号	组件类型	属性名称	属性值	说明
1	TextView	android:id	@+id/tv_title	增加 id 为"tv_title"的组件
		android:text	梅花布局效果	显示文本为"梅花布局效果"
		android:layout_centerHorizontal	true	水平方向居中对齐
		android:layout_alignParentTop	true	与父容器顶端对齐
		android:textSize	20sp	文本大小为 20sp
		android:textColor	#ff0080	文本颜色为#ff0080
2	TextView	android:id	@+id/tv_center	增加 id 为"tv_center"的组件
		android:text	进入	显示文本为"进入"
		android:background	@drawable/bg1	文本框的背景图片为 res\drawable\bg1
		android:layout_centerInParent	true	组件在父容器中居中
3	TextView	android:id	@+id/tv_top	增加 id 为"tv_top"的组件
		android:text	帮助	显示文本为"帮助"
		android:background	@drawable/bg3	文本框的背景图片为 res\drawable\bg3
		android:layout_above	@id/tv_center	位于 tv_center 的上方
		android:layout_alignLeft	@id/tv_center	与 tv_center 左对齐
4	TextView	android:id	@+id/tv_bottom	增加 id 为"tv_bottom"的组件
		android:text	联络	显示文本为"联络"
		android:background	@drawable/bg3	文本框的背景图片为 res\drawable\bg3
		android:layout_below	@id/tv_center	位于 tv_center 的下方
		android:layout_alignLeft	@id/tv_center	与 tv_center 左对齐
5	TextView	android:id	@+id/tv_left	增加 id 为"tv_left"的组件
		android:text	设置	显示文本为"设置"
		android:background	@drawable/bg3	文本框的背景图片为 res\drawable\bg3
		android:layout_toLeftOf	@id/tv_center	位于 tv_center 的左侧
		android:layout_alignTop	@id/tv_center	与 tv_center 顶端对齐
6	TextView	android:id	@+id/tv_right	增加 id 为"tv_right"的组件
		android:text	退出	显示文本为"退出"
		android:background	@drawable/bg3	文本框的背景图片为 res\drawable\bg3
		android:layout_toRightOf	@id/tv_center	位于 tv_center 的右侧
		android:layout_alignTop	@id/tv_center	与 tv_center 顶端对齐

参考代码如下：

```xml
<!--定义组件 -->
<!--定义标题文本 -->
<TextView
    android:id = "@+id/tv_title"
    android:layout_width = "wrap_content"
    android:layout_height = "wrap_content"
    android:text = "梅花布局效果"
    android:layout_centerHorizontal = "true"
    android:layout_alignParentTop = "true"
    android:textSize = "20sp"
    android:textColor = "#ff0080"/>
<!--定义位于父容器中间的组件 -->
<TextView
    android:id = "@+id/tv_center"
    android:layout_width = "wrap_content"
    android:layout_height = "wrap_content"
    android:text = "进入"
    android:background = "@drawable/bg1"
    android:layout_centerInParent = "true"/>
<!--定义位于tv_center上方且与其左对齐的组件 -->
<TextView
    android:id = "@+id/tv_top"
    android:layout_width = "wrap_content"
    android:layout_height = "wrap_content"
    android:text = "帮助"
    android:background = "@drawable/bg3"
    android:layout_above = "@id/tv_center"
    android:layout_alignLeft = "@id/tv_center"/>
<!--定义位于tv_center组件下方且与其左对齐的组件 -->
<TextView
    android:id = "@+id/tv_bottom"
    android:layout_width = "wrap_content"
    android:layout_height = "wrap_content"
    android:text = "联络"
    android:background = "@drawable/bg3"
    android:layout_below = "@id/tv_center"
    android:layout_alignLeft = "@id/tv_center"/>
<!--定义位于tv_center组件左侧且与其顶端对齐的组件 -->
<TextView
    android:id = "@+id/tv_left"
    android:layout_width = "wrap_content"
    android:layout_height = "wrap_content"
    android:text = "设置"
    android:background = "@drawable/bg3"
    android:layout_toLeftOf = "@id/tv_center"
    android:layout_alignTop = "@id/tv_center"/>
```

```xml
<!--定义位于 tv_center 组件右侧且与其顶端对齐的组件-->
<TextView
    android:id = "@+id/tv_right"
    android:layout_width = "wrap_content"
    android:layout_height = "wrap_content"
    android:text = "退出"
    android:background = "@drawable/bg3"
    android:layout_toRightOf = "@id/tv_center"
    android:layout_alignTop = "@id/tv_center"/>
```

2.6.2 相对布局（RelativeLayout）的定义与使用

相对布局是布局最常用，也是最灵活的一种布局，用 RelativeLayout 类来表示。在相对布局中组件的位置可以参考其他组件的位置来确定。只有此布局中的各组件与位置相关的属性才会生效，如 android:layout_below、android:layout_above 等。子元素就通过这些属性和各自的 ID 配合指定位置关系。

需要注意的是，在指定位置关系时，引用的 ID 必须在引用之前先被定义，否则将出现异常。这就要求在相对布局时，如果 A 组件的位置是由 B 组件的位置来决定，则要先定义 B 组件，再定义 A 组件。

RelativeLayout 的常用 XML 属性及相关说明见表 2-15。

表 2-15 RelativeLayout 的常用 XML 属性及相关说明

序号	XML 属性	说明	编码举例
1	android:gravity	设置该布局容器内部各子组件的对齐方式 可选值：top、bottom、left、right、center_vertical、center_horizontal、center、fill、clip_vertical 和 clip_horizontal	android:gravity = "top|left"，表示将布局内子组件置于左上角
2	android:ignoreGravity	设置哪个组件不受 gravity 属性的影响 值为组件 ID	android:ignoreGravity = "@+id/btnOK"，表示 id 为 btnOK 的组件不受 gravity 属性的影响

RelativeLayout 中组件与其他组件位置相关的常用 XML 属性和相关说明见表 2-16。

表 2-16 RelativeLayout 中组件与其他组件位置有关的常用 XML 属性和相关说明

序号	XML 属性	说明	编码举例
1	android:layout_above	摆放在指定 ID 组件的上方	android:layout_above = "@id/tv_center"，表示将组件摆放到 ID 为 tv_center 的上方
2	android:layout_below	摆放在指定 ID 组件的下方	android:layout_below = "@id/tv_center"，表示将组件摆放到 ID 为 tv_center 的下方
3	android:layout_toLeftOf	摆放在指定 ID 组件的左侧	android:layout_toLeftOf = "@id/tv_center"，表示将组件摆放到 ID 为 tv_center 的左侧
4	android:layout_toRightOf	摆放在指定 ID 组件的右侧	android:layout_toRightOf = "@id/tv_center"，表示将组件摆放到 ID 为 tv_center 的右侧
5	android:layout_alignTop	以指定 ID 组件为参考上对齐	android:layout_alignTop = "@id/tv_center"，表示将组件与 ID 为 tv_center 的组件上对齐

(续)

序号	XML 属性	说明	编码举例
6	android:layout_alignBottom	以指定 ID 组件为参考下对齐	android:layout_alignBottom = "@id/tv_center"，表示将组件与 ID 为 tv_center 的组件下对齐
7	android:layout_alignLeft	以指定 ID 组件为参考左对齐	android:layout_alignLeft = "@id/tv_center"，表示将组件与 ID 为 tv_center 的组件左对齐
8	android:layout_alignRight	以指定 ID 组件为参考右对齐	android:layout_alignRight = "@id/tv_center"，表示将组件与 ID 为 tv_center 的组件右对齐
9	注意：上述属性的值均为指定组件的 ID，可以通过"R.id.组件的 ID 值"的方式来访问		

RelativeLayout 中组件与父容器的位置关系对应的常用 XML 属性和相关说明见表 2-17。

表 2-17　RelativeLayout 中组件与父容器位置有关的常用 XML 属性和相关说明

序号	XML 属性	说明	编码举例
1	android:layout_alignParentLeft	该组件是否对齐父组件的左端	android:layout_alignParentLeft = "true"，表示组件与父组件左对齐
2	android:layout_alignParentRight	该组件是否对齐父组件的右端	android:layout_alignParentRight = "true"，表示组件与父组件右对齐
3	android:layout_alignParentTop	该组件是否对齐父组件的顶部	android:layout_alignParentTop = "true"，表示组件与父组件顶端对齐
4	android:layout_alignParentBottom	该组件是否对齐父组件的底部	android:layout_alignParentBottom = "true"，表示组件与父组件底部对齐
5	android:layout_centerInParent	该组件是否相对于父组件居中	android:layout_centerInParent = "true"，表示组件相对于父组件居中对齐
6	android:layout_centerHorizontal	该组件是否横向居中	android:layout_centerHorizontal = "true"，表示组件横向居中
7	android:layout_centerVertical	该组件是否垂直居中	android:layout_centerVertical = "true"，表示组件垂直居中
8	注意：上述属性的值均为逻辑值 true 或 false		

2.7　实例 7：DIY 计算器

2.7.1　功能要求与操作步骤

1. 功能要求

完成如图 2-14 所示的"计算器"应用程序界面。要求能够：

① 计算器由 4 列组成，在屏幕上居中显示。
② 左上角显示运算结果，数字键和运算键合理放置。
③ 其他：数字键"00"跨 2 列显示，"="跨 3 列显示，运算键"+"跨 3 行显示。

2. 操作步骤

（1）创建项目

参考如下输入信息创建 Android 项目：

图 2-14　网格布局的应用示例图

> Application Name：07_TestGridLayout
> Project Name：07_TestGridLayout
> Package Name：com.book.testgridlayout
> Activity Name：MainActivity
> Layout Name：activity_main

（2）设置布局管理器属性

参考表 2-18 设置其属性。

表 2-18 实例 7 中 GridLayout 的属性列表

序 号	属性名称	属 性 值	说 明
1	android：layout_width	wrap_content	宽度适应内容即可
2	android：layout_height	wrap_content	高度适应内容即可
3	android：columnCount	4	网格中包含 4 列
4	android：layout_gravity	center	在屏幕上居中显示网格布局

参考代码如下：

```
<GridLayout xmlns:android="http://schemas.android.com/apk/res/android"
    android:layout_width="wrap_content"
    android:layout_height="wrap_content"
    android:columnCount="4"
    android:layout_gravity="center"
>
    <!--定义组件-->
    …
</GridLayout>
```

（3）添加组件

从图 2-14 可看出在此布局中，至少需要 1 个 TextView（或 EditText）和 16 个 Button。其宽度和高度属性均可设置为"wrap_content"，其余属性可参考表 2-19 设置。

表 2-19 实例 7 各组件的属性列表

序号	组件类型	属性名称	属 性 值	说 明
1	TextView	android：id	@+id/textView0	增加 id 为"textView0"的组件
		android：text	运算结果显示区	显示文本为"运算结果显示区"
		android：layout_columnSpan	3	组件跨 3 列显示
		android：layout_gravity	fill	组件以填充方式占满 3 列
		android：gravity	center	文本在组件中居中对齐
2	Button（数字 0-9、除号"/"、减号"-"和乘号"*"）	android：text	相应数字或值，如"0"	设置按钮的显示文本为"0"（或其他）

（续）

序号	组件类型	属性名称	属性值	说明
3	Button（加号"+"）	android:text	+	显示文本为"+"
		android:layout_rowSpan	3	组件跨3行显示
		android:layout_gravity	fill	组件以填充方式占满3行
4	Button（数字"00"）	android:text	00	显示文本为"00"
		android:layout_columnSpan	2	组件跨2列显示
		android:layout_gravity	fill	组件以填充方式占满2列
5	Button（等号"="）	android:text	=	显示文本为"="
		android:layout_columnSpan	3	组件跨3列显示
		android:layout_gravity	fill	组件以填充方式占满3列

按照上述网格布局的定义方式，需要按照逐行、从左到右的方式定义各个组件；也可以为每个组件通过设置其 android:layout_row、android:layout_column 指定所在行、列的编号分别进行定义。参考代码如下：

```
<!--定义组件-->
<TextView
    android:id = "@+id/textView1"
    android:layout_gravity = "fill"
    android:layout_columnSpan = "3"
    android:text = "运算结果显示区"
    android:gravity = "center"
/>
<Button android:text = "/" />

<Button android:text = "1"/>

<Button android:text = "2"/>

<Button android:text = "3"/>

<Button android:text = " * "/>

<Button android:text = "4"/>

<Button android:text = "5"/>

<Button android:text = "6"/>

<Button android:text = " - "/>
```

```
        < Button android:text = "7"/>

        < Button android:text = "8"/>

        < Button android:text = "9"/>

        < Button
            android:text = " + "
            android:layout_gravity = "fill"
            android:layout_rowSpan = "3"
            />

        < Button android:text = "0"/>

        < Button
            android:text = "00"
            android:layout_columnSpan = "2"
            android:layout_gravity = "fill"/>

        < Button
            android:text = " = "
            android:layout_columnSpan = "3"
            android:layout_gravity = "fill"
            />
```

2.7.2 网格布局（GridLayout）的定义与使用

如上文所述，使用频率最高的布局是 LinearLayout，它可以让布局界面中的子控件以常见的方式（如水平或者垂直方向）对齐。在使用 LinearLayout 时，经常为了达到理想的布局效果，去使用各种 LinearLayout 进行嵌套。然而在嵌套使用布局时，应用程序会产生性能低下等问题。为了解决这些问题，在 Android 4.0 及 4.0 的以上版本中，新增加了网格布局，用 GridLayout 类来表示。

GridLayout 布局将子组件放在一个矩形的网格中。具体地，它使用虚细线将布局划分为行、列和单元格，也支持一个控件在行、列上都有交错排列。布局中的网格线可以通过访问其下标来取得。例如，在一个有 N 列的网格布局中，无论如何配置布局，分隔线的索引值都从 0 开始到 N 结束，共 N + 1 条。

在使用 GridLayout 时，需要在配置项目时将 minSdkVersion 设置为 14。在配置文件 AndroidManifest.xml 文件中体现为如下代码：

```
        < uses – sdk
            android:minSdkVersion = "14"
            android:targetSdkVersion = "17"/>
```

GridLayout 的常用 XML 属性和相关说明见表 2-20 所示。

表 2-20 GridLayout 的常用 XML 属性和相关说明

序号	XML 属性	说 明	编码举例
1	android:alignmentMode	设置网格内部组件的对齐方式 可选值：alignBounds（按边界对齐）和 alignMargins（按边距对齐）	android:alignmentMode = "alignMargins"，表示网格内部组件按照边距值对齐
2	android:columnCount	设置网格的最大列数	android:columnCount = "3"，表示当自动定位子组件时最大列数为 3 列
3	android:rowCount	设置网格的最大行数	android:rowCount = "3"，表示当自动定位子组件时最大行数为 3 列

2.8 实例 8：美食背后的故事

2.8.1 功能要求与操作步骤

在许多应用中，图文并茂、主要功能在同页面上合理共存是一个基本需求。然而，单纯使用一种布局方式达到这个目的却不是件容易的事情。本节将通过菜谱应用的首页设计介绍如何嵌套使用各种布局。

1. 功能要求

完成如图 2-15 所示的菜谱应用程序界面。要求能够：

① 如图 2-15a，在屏幕最上方居中显示美食相关图片。

② 在图片下方显示美食的名称和美食背后的故事，文字较多时可以通过上下滑动屏幕看到全部文字，效果如图 2-15b 所示。

③ 在屏幕下方的左、右两侧分别居中显示两个图形按钮。

a)　　　　　　　　　　　b)

图 2-15　嵌套布局的应用示例图

2. 操作步骤

（1）创建项目

参考如下输入信息创建 Android 项目：

> Application Name：08_TestNestLayout
> Project Name：08_TestNestLayout
> Package Name：com. book. testnestlayout
> Activity Name：MainActivity
> Layout Name：activity_main

（2）修改布局管理器

从图 2-15 可以看出，应用程序界面在整体上可以分为可以滚动的图文和按钮两部分。上半部分是由图片、两个文本竖直方向线性排列，下半部分的两个按钮则按水平方向线性排列。为了允许滚动浏览较长文本，需要在上半部分的线性布局外层套一个 ScrollView。为了使两个按钮能分别居中旋转在水平线性布局的左、右两侧，需要在下半部分的线性布局内部增加两个线性布局，并将其 android:gravity 属性设置为"center"。

ScrollView 即滚动视图，允许通过滚动条来显示在屏幕不能显示完整的内容。它是 FrameLayout 的一个子类，这就意味着需要在其中放置有滚动内容的子元素。子元素可以是一个复杂的对象的布局管理器，但通常使用的子元素是垂直方向的 LinearLayout。需要注意的是 ScrollView 只支持垂直滚动。

其整体布局的框架结构如图 2-16 所示。

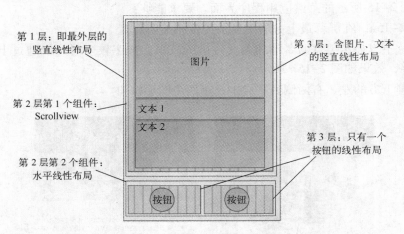

图 2-16 嵌套布局的应用示例图

可参考表 2-21 设置各个布局及组件的属性。

表 2-21 实例 8 中各组件的属性列表

序号	组件类型	属性名称	属性值	说明
1	最外层的 LinearLayout	android:layout_width	match_parent	宽度匹配父容器
		android:layout_height	match_parent	高度匹配父容器
		android:orientation	vertical	竖直方向

（续）

序号	组件类型	属性名称	属性值	说明
2	ScrollView	android:layout_width	match_parent	宽度匹配父容器
		android:layout_height	0dp	与 android:layout_weight 属性配合使用,按权重分配高度
		android:layout_weight	8	高度在父容器中所占比例为80%
3	嵌套在 ScrollView 中的 LinearLayout	android:layout_width	match_parent	宽度匹配父容器
		android:layout_height	match_parent	高度匹配父容器
		android:orientation	vertical	竖直方向
4	ImageView(图片)	android:layout_width	match_parent	宽度匹配父容器
		android:layout_height	wrap_content	高度适合内容即可
		android:src	@drawable/gbjd	显示 res/drawable/gbjd 图片
5	TextView(文本1)	android:layout_width	match_parent	宽度匹配父容器
		android:layout_heigh	wrap_content	高度适合内容即可
		android:text	@string/gbjdbt	显示文本为 string/gbjdbt 中的内容
		android:textAppearance	?android:attr/textAppearanceLarge	大文本样式
		android:textColor	#ff7800	设置文本颜色
6	TextView(文本2)	android:layout_width	match_parent	宽度匹配父容器
		android:layout_heigh	wrap_content	高度适合内容即可
		android:text	@string/gbjd	显示文本为 string/gbjd 中的内容
7	第2层中的水平线性布局 LinearLayout	android:layout_width	match_parent	宽度匹配父容器
		android:layout_height	0dp	与 android:layout_weight 属性配合使用,按权重分配高度
		android:layout_weight	2	高度在父容器中所占比例为20%
8	两个放按钮的线性布局 LinearLayout	android:layout_width	0dp	与 android:layout_weight 属性配合使用,按权重分配高度
		android:layout_height	match_parent	高度匹配父容器
		android:layout_weight	1	宽度在父容器中所占比例均为50%
		android:gravity	center	组件在其中居中对齐
9	两个按钮 ImageButton	android:layout_width	wrap_content	宽度适合内容即可
		android:layout_height	match_parent	高度匹配父容器
		android:src	左侧按钮:@drawable/zan 右侧按钮:@drawable/cai	左侧按钮显示图片为:res/drawable/zan 右侧按钮显示图片为:res/drawable/cai
		android:background	#fff	设置按钮的背景色

整个布局的参考代码如下:

```
<!-- 第1层:竖直线性布局 -->
< LinearLayout xmlns:android = "http://schemas.android.com/apk/res/android"
```

```xml
        android:layout_width = "match_parent"
        android:layout_height = "match_parent"
        android:orientation = "vertical" >
<!-- 第2层:上半部分的滚动视图 -->
<ScrollView
        android:layout_width = "match_parent"
        android:layout_height = "0dp"
        android:layout_weight = "8" >
<!-- 第3层:包含图片和两个文本的竖直线性布局 -->
<LinearLayout
        android:layout_width = "match_parent"
        android:layout_height = "match_parent"
        android:orientation = "vertical" >
<!-- 图片 -->
<ImageView
        android:layout_width = "match_parent"
        android:layout_height = "wrap_content"
        android:src = "@drawable/gbjd"/>
<!-- 文本1 -->
<TextView
        android:layout_width = "match_parent"
        android:layout_height = "wrap_content"
        android:text = "@string/gbjdbt"
        android:textAppearance = "?android:attr/textAppearanceLarge"
        android:textColor = "#ff7800"/>
<!-- 文本2 -->
<TextView
        android:layout_width = "match_parent"
        android:layout_height = "wrap_content"
        android:text = "@string/gbjd"/>
</LinearLayout>
</ScrollView>
<!-- 第2层:下半部分的水平线性布局 -->
    <LinearLayout
        android:layout_width = "match_parent"
        android:layout_height = "0dp"
        android:layout_weight = "2" >
<!-- 第3层:放左侧按钮的线性布局 -->
<LinearLayout
        android:layout_width = "0dp"
        android:layout_height = "match_parent"
        android:layout_weight = "1"
        android:gravity = "center" >
<!-- 左侧按钮 -->
<ImageButton
        android:layout_width = "wrap_content"
        android:layout_height = "match_parent"
        android:src = "@drawable/zan"
```

```
                    android:background = "#fff"/>
        </LinearLayout>
        <!-- 第3层:放右侧按钮的线性布局 -->
        <LinearLayout
                    android:layout_width = "0dp"
                    android:layout_height = "match_parent"
                    android:layout_weight = "1"
                    android:gravity = "center" >
            <!-- 右侧按钮 -->
            <ImageButton
                    android:layout_width = "wrap_content"
                    android:layout_height = "match_parent"
                    android:background = "#fff"
                    android:src = "@drawable/cai"/>
        </LinearLayout>
    </LinearLayout>
</LinearLayout>
```

本例布局中使用的两个字符串资源必须在 res/values/strings.xml 中被定义,参考代码如下:

```
<?xml version = "1.0" encoding = "utf-8"?>
<resources>
    ...
    <string name = "gbjd" >\t\t宫保鸡丁是…后风靡全国直至今日。</string>
    <string name = "gbjdbt" >宫保鸡丁</string>
</resources>
```

2.8.2 布局的嵌套使用

经验告诉开发者,在 Android 开发中 UI 设计十分重要,当用户使用一个软件时,最先感受到的不是这款软件的功能是否强大,而是界面设计是否精致,用户体验是否良好。Android 中的 UI 是由布局和组件协同完成的,布局如同是一幢建筑中的框架结构,而组件相当于一幢建筑中的砖瓦。布局可以定义为组件在 UI 中的呈现方式,即组件大小、间距和对齐方式等。

在 Android 中提供了以下两种创建布局的方式。

第一,在 XML 配置文件中声明,这种方式是将需要呈现的组件在配置文件中进行声明。它通常保存在 res\layout\文件夹下,文件名任意,但需要以英文字母开头且不能含大写字母。

在 Java 中,可以通过调用 Activity 类的 setContentView(int layoutResID)方法,将布局文件展示在 Activity 中。setContentView 方法的参数是布局文件资源 ID,需要用"R.layout.布局文件名称"的方法来取得。例如,定义的布局文件名称是"main.xml",在 MainActivity.java 文件的 onCreate()方法中就可以通过如下代码使用此布局:

```
setContentView(R.layout.main)
```

在使用布局中的组件时，可以调用 Activity 类的 findViewById(int id) 方法，其参数是组件 id，需要用"R. id. 组件 id"的方法来取得。例如以下代码即可将 XML 文件中的 id 为 "linearLayout"的布局组件找出，并赋值于 View 类的对象 linear_layout：

> linear_layout = findViewById(R. id. linearLayout)

第二，在 Java 中通过"硬代码"直接实例化布局及其组件。

由于 Android 平台在一定程度上更面向对象，在构架上仍然使用 MVC 这样的 UI 和代码逻辑分离的思路，以便 UI 可以专心地设计用户界面，代码复用和维护性也可以得到提高。所以在本书中，推荐使用第 1 种方式实现布局 UI。

在 5 种布局中，相对布局、线性布局和表格布局的使用频率较高。在使用布局管理器进行组件布局时，这 5 种布局既可以单独使用，也可以嵌套使用，在实际应用中应根据需要灵活安排。

2.9 实例 9：简易文本阅读器

2.9.1 功能要求与操作步骤

1. 功能要求

完成如图 2-17 所示的"简易文本阅读器"应用程序。要求能够：

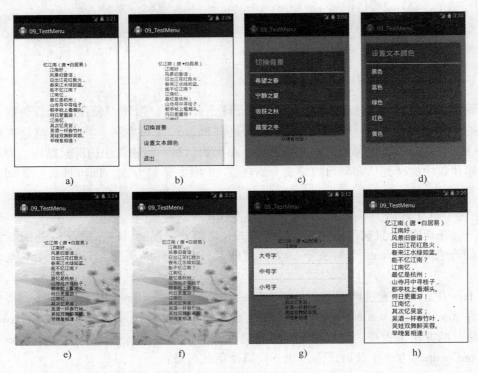

图 2-17 简易文本阅读器运行效果

① 应用程序的主界面如图2-17a所示。

② 在运行应用程序后,当用户按下"Menu"键,则弹出如图2-17b所示的包含"切换背景"、"设置文本颜色"和"退出"选项菜单。

③ 在用户选择"切换背景"和"设置文本颜色"菜单后,分别弹出如图2-17c和图2-17d所示的子菜单。

④ 在用户选择子菜单项后,应用程序做出相应设置。如选择"切换背景"中的子菜单项"宁静之夏",则应用程序效果如图2-17e。如果再选择"设置文本颜色"中的"蓝色",则应用程序效果如图2-17f所示。

⑤ 常按应用程序中的"文本",会弹出如图2-17g所示的上下文菜单,各菜单项用以调整文本框中文本的字号大小。图2-17h展示了用户选择"大号字"菜单后的运行效果。

2. 操作步骤

(1) 创建项目

参考如下输入信息创建Android项目:

```
Application Name:09_TestMenu
Project Name:09_TestMenu
Package Name:com.book.testmenu
Activity Name:MainActivity
Layout Name:activity_main
```

(2) 准备图片和字符串资源

在应用程序中,需要将4幅用来设置应用程序背景的图片分别粘贴到项目包中的"res\drawable"目录下,其名称分别是"background_02_spring""background_02_summer""background_02_autumn"和"background_02_winter"。

为了方便使用、统一更改,各个菜单项的标题以及其他文本可以预先在"res\values\strings.xml"中定义。参考代码如下:

```xml
<?xml version="1.0" encoding="utf-8"?>
<resources>
    <!-- 项目名称 -->
<string name="app_name">09_TestMenu</string>
<!-- 各种菜单标题 -->
    <string name="changeBackground">切换背景</string>
    <string name="background_spring">希望之春</string>
    <string name="background_summer">宁静之夏</string>
    <string name="background_autumn">收获之秋</string>
    <string name="background_winter">晶莹之冬</string>

    <string name="changeSize">调整字号</string>
    <string name="size_big">大号字</string>
    <string name="size_mid">中号字</string>
    <string name="size_small">小号字</string>


    <string name="changeColor">设置文本颜色</string>
    <string name="text_color_black">黑色</string>
```

```xml
<string name="text_color_blue">蓝色</string>
<string name="text_color_green">绿色</string>
<string name="text_color_red">红色</string>
<string name="text_color_yellow">黄色</string>

<string name="exit">退出</string>

<!-- 显示文本 -->
<string name="text_content">
忆江南(唐·白居易)
\n\t 江南好,
    \n\t 风景旧曾谙;
    \n\t 日出江花红胜火,
    \n\t 春来江水绿如蓝。
    \n\t 能不忆江南?

    \n\t 江南忆,
    \n\t 最忆是杭州;
    \n\t 山寺月中寻桂子,
    \n\t 郡亭枕上看潮头。
    \n\t 何日更重游!
...
</string>
</resources>
```

(3) 设计菜单

如前所述,在"res/menu"中创建两个菜单文件,名称分别是"main_menu.xml"和"context_menu.xml"。

"main_menu.xml"用作应用程序的选项菜单,参考代码如下:

```xml
<menu xmlns:android="http://schemas.android.com/apk/res/android">
    <!--选项菜单:切换背景-->
<item
        android:id="@+id/changeBackground"
        android:title="@string/changeBackground">
    <!--切换背景子菜单-->
<menu>
    <item
        android:id="@+id/background_spring"
        android:title="@string/background_spring"/>
    <item
        android:id="@+id/background_summer"
        android:title="@string/background_summer"></item>
    <item
        android:id="@+id/background_autumn"
        android:title="@string/background_autumn"/>
    <item
```

```xml
                android:id = "@+id/background_winter"
                android:title = "@string/background_winter"/>
        </menu>
    </item>
    <!-- 选项菜单:设置文本颜色 -->
    <item
                android:id = "@+id/changeColor"
                android:title = "@string/changeColor">
        <!-- 设置文本颜色子菜单 -->
        <menu>
        <item
                android:id = "@+id/text_color_black"
                android:title = "@string/text_color_black"/>
        <item
                android:id = "@+id/text_color_blue"
                android:title = "@string/text_color_blue"/>
        <item
                android:id = "@+id/text_color_green"
                android:title = "@string/text_color_green"/>
        <item
                android:id = "@+id/text_color_red"
                android:title = "@string/text_color_red"/>
        <item
                android:id = "@+id/text_color_yellow"
                android:title = "@string/text_color_yellow"/>
        </menu>
    </item>
    <!-- 选项菜单:退出 -->
    <item
            android:id = "@+id/exit"
            android:title = "@string/exit"/>
</menu>
```

"context_menu.xml"用作文本框的上下文菜单,参考代码如下:

```xml
<?xml version = "1.0" encoding = "utf-8"?>
<menu xmlns:android = "http://schemas.android.com/apk/res/android">
    <!-- 上下文菜单:调整字号 -->
    <item
            android:id = "@+id/size_big"
            android:title = "@string/size_big"/>
    <item
            android:id = "@+id/size_mid"
            android:title = "@string/size_mid"/>
    <item
            android:id = "@+id/size_small"
            android:title = "@string/size_small"/>
</menu>
```

（4）修改布局管理器

在这个应用程序的主界面中只有一个组件——TextView，所以可以将默认的布局管理器"RelativeLayout"更改为"FrameLayout"。可参考表2-22设置布局及TextView组件的属性。

表2-22　实例9中各组件的属性列表

序号	组件类型	属性名称	属性值	说明
1	FrameLayout	android:layout_width	match_parent	宽度匹配父容器
		android:layout_height	match_parent	高度匹配父容器
		android:id	@+id/bg	增加id为"bg"的组件
2	TextView	android:layout_width	wrap_content	宽度适合内容即可
		android:layout_height	wrap_content	高度适合内容即可
		android:id	@+id/tv_content	增加id为"tv_content"的组件
		android:layout_gravity	center	居中对齐
		android:text	@string/text_content	设置文本为string中text_content定义的内容

整个布局的参考代码如下：

```xml
<FrameLayout xmlns:android="http://schemas.android.com/apk/res/android"
    android:layout_width="match_parent"
    android:layout_height="match_parent"
    android:id="@+id/bg"
    >
    <TextView
        android:id="@+id/tv_content"
        android:layout_width="wrap_content"
        android:layout_height="wrap_content"
        android:text="@string/text_content"
        android:layout_gravity="center"/>
</FrameLayout>
```

（5）实现MainActivity中的功能

```java
public class MainActivity extends Activity {
    //变量声明
    FrameLayout layout;
    Drawable background;
    TextView tv;
    float ini_text_size,changed_text_size;
    int text_color;
    @Override
    protected void onCreate(Bundle savedInstanceState) {
        super.onCreate(savedInstanceState);
        //使用activity_main作为当前activity的布局
        setContentView(R.layout.activity_main);
        //找出相应组件
        layout = (FrameLayout) this.findViewById(R.id.bg);
```

```java
            tv = (TextView) this.findViewById(R.id.tv_content);
            //取得文本原有字号大小和颜色
            ini_text_size = tv.getTextSize();
            changed_text_size = ini_text_size;
            text_color = Color.BLACK;
            //为文本框注册上下文菜单
            this.registerForContextMenu(tv);
    }
    //创建上下文菜单
    @Override
    public void onCreateContextMenu(ContextMenu menu, View v, ContextMenuInfo menuInfo) {
            getMenuInflater().inflate(R.menu.context_menu, menu);
            super.onCreateContextMenu(menu, v, menuInfo);
    }
    //创建选项菜单
    @Override
    public boolean onCreateOptionsMenu(Menu menu) {
            // Inflate the menu; this adds items to the action bar if it is present.
            getMenuInflater().inflate(R.menu.main_menu, menu);
            return true;
    }
    //实现各选项菜单项被选中时的事件
    @Override
    public boolean onOptionsItemSelected(MenuItem item) {
            //通过选项菜单项的id判断哪个菜单项被选中
            switch(item.getItemId()) {
            case R.id.background_autumn:
                background = getResources().getDrawable(R.drawable.background_02_autumn);
                break;
            case R.id.background_spring:
                background = getResources().getDrawable(R.drawable.background_02_spring);
                break;
            case R.id.background_summer:
                background = getResources().getDrawable(R.drawable.background_02_summer);
                break;
            case R.id.background_winter:
                background = getResources().getDrawable(R.drawable.background_02_winter);
                break;
            case R.id.text_color_black:
                text_color = Color.BLACK;
                break;
            case R.id.text_color_blue:
                text_color = Color.BLUE;
                break;
            case R.id.text_color_green:
                text_color = Color.GREEN;
                break;
            case R.id.text_color_red:
```

```
            text_color = Color. RED;
            break;
        case R. id. text_color_yellow:
            text_color = Color. YELLOW;
            break;
        case R. id. exit:
            System. exit(0);
            break;
        }
        //设置背景
        layout. setBackground(background);
        //设置文本颜色
        tv. setTextColor(text_color);
        return super. onOptionsItemSelected(item);
    }
    //实现上下文菜单项被选中时的事件
    @Override
    public boolean onContextItemSelected(MenuItem item) {
        //通过上下文菜单项的 ID 值判断哪个上下文菜单被选中
        switch(item. getItemId()) {
        case R. id. size_big:
            changed_text_size = ini_text_size + 5;
            break;
        case R. id. size_mid:
            changed_text_size = ini_text_size;
            break;
        case R. id. size_small:
            changed_text_size = ini_text_size - 5;
            break;
        }
        //设置文本的字号大小为 changed_text_size
        tv. setTextSize(changed_text_size);
        return super. onContextItemSelected(item);
    }
}
```

2.9.2 菜单（Menu）资源的定义与使用

为了使应用程序有更完善的用户体验，除了设计人性化的用户界面外，添加一些菜单也是必要的。Android 中的菜单分为选项菜单、上下文菜单和子菜单。这些菜单都可以在 res/menu 中以 XML 的格式声明定义，通过调用"MenuInflater"类的"inflate"（R. menu. 菜单名称）的方式使用。

选项菜单是一种最基本的菜单。通过单击 Android 手机上的"Menu"键，即可看到选项菜单，它通常在屏幕的底部显示。在 Android 4.2 以上版本的手机中，菜单不再像以前版本中一样横向显示，而变为纵向显示。通过覆写 Activity 类的"onCreateOptionsMenu（Menu menu）"方法可以创建选项菜单。

上下文菜单类似于 Windows 操作系统中的右键菜单。在 Android 应用中，如果为某个组件注册了上下文菜单，那么当用户长按该组件时会看到这个菜单。通过覆写 Activity 类的"onCreateContextMenu(Menu menu, View v, ContextMenu ContextMenuInfo)"方法可以创建上下文菜单。

当在系统中定义完菜单之后，也可以为每个菜单定义多个子菜单。

在 res/menu 当中的 XML 文件中，添加菜单的方法如下：

```xml
< menu xmlns:android = "http://schemas.android.com/apk/res/android" >
    <!-- 菜单项 -->
    < item
            android:id = "@+id/菜单 ID"
            android:title = "菜单标题" >
        <!-- 子菜单 -->
        < menu >
            < item
                    android:id = "@+id/子菜单 ID"
                    android:title = "子菜单标题"/>
            < item
                    android:id = "@+id/子菜单 ID"
                    android:title = "子菜单标题"/>
            ...
        </menu>
    </item>
    <!-- 菜单项 -->
    < item
            android:id = "@+id/菜单 ID"
            android:title = "菜单标题" >
    </item>
    ...
</menu>
```

2.10 动手实践 2：紫禁城一日游

2.10.1 功能要求

1. 应用程序首界面

应用程序的首界面由上下两部分组成，上半部分显示紫禁城的标志图，下半部分是可以上下滚动的文字区域，默认显示紫禁城的简介信息，如图 2-18 所示。

2. 为应用添加选项菜单

当用户单击了"Menu"键后，可以弹出如图 2-19a 所示的选项菜单，其中"餐饮"菜单下有子菜单"北京烤鸭""涮羊肉""满汉全席""风味小吃"，单击各个菜单项可以在文字区域中显示相应内容，效果如图 2-19b 所示。

3. 为文字添加上下文菜单

在文字区域中长按时，会弹出调整字号的上下文菜单"调大字号"和"调小字号"，效果如图 2-20a。单击菜单项，分别会将文字区域中的文字变大或缩小，图 2-20b 展示了调大字号的效果。

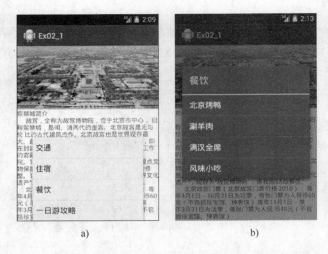

图 2-18 "紫禁城一日游"首界面　　　　　图 2-19 "紫禁城一日游"菜单效果

a) 　　　　　　　　　b)

图 2-20 调整字号菜单效果

2.10.2 操作提示

1. 定义需要的布局

应用程序的首界面中展示图片的部分需要用到 ImageView 组件，显示文字的部分需要用到 TextView 组件。为了使文字可以上下滚动，需要用到 ScrollView 组件。布局参考代码如下：

```
< LinearLayout xmlns:android = "http://schemas.android.com/apk/res/android"
    android:layout_width = "fill_parent"
    android:layout_height = "fill_parent"
```

```
            android:orientation = "vertical" >
    <ImageView
            android:id = "@+id/img"
            android:layout_width = "match_parent"
            android:layout_height = "wrap_content"
            android:src = "@drawable/forbiddencity"
            android:scaleType = "fitXY"/>
    <ScrollView
        android:layout_width = "wrap_content"
            android:layout_height = "wrap_content" >
    <LinearLayout
            android:layout_width = "wrap_content"
android:layout_height = "wrap_content" >
            <TextView
                    android:id = "@+id/tv"
            android:layout_width = "wrap_content"
            android:layout_height = "wrap_content"
            android:text = "@string/introduction"/>
    </LinearLayout >
    </ScrollView >
</LinearLayout >
```

2. 制作菜单

读者可以参考第2.9节内容完成选项菜单、子菜单和上下文菜单的制作。

第3章　Android中的基本视图组件

在许多高级程序设计语言中，都提供了图形用户界面（Graphical User Interface，GUI）以实现人机交互的操作。在Android中也同样提供了大量的显示组件，如第2章中大量使用的TextView，以及本章即将介绍的Button等，这些都是Android应用开发中经常使用到的UI视图组件。使用合理的布局管理器，配合合适的UI组件可以让界面显得更加丰富多彩。为了让UI组件可以响应用户的不同操作，还需要为其添加相应的事件监听，并编写相应的事件处理代码。

3.1　实例1：新闻摘要与详情

3.1.1　功能要求与操作步骤

1. 功能要求

完成如图3-1所示的应用程序。要求能够：

① 在图3-1a所示的应用程序首页中，新闻标题使用如下样式参数：字号为20sp，文字颜色为蓝色，加粗显示；新闻内容使用如下样式参数：字号为15sp，文字颜色为黑色。

② 单击首页右下角的"详情"，可以打开新闻的相关链接，如图3-1b所示。

图3-1　"新闻摘要与详情"运行效果图

2. 操作步骤

（1）创建项目

参考如下输入信息创建Android项目：

```
Application Name: 01_TestTextView
Project Name: 01_TestTextView
Package Name: com.book.testtextview
Activity Name: MainActivity
Layout Name: activity_main
```

(2) 准备字符串资源

为了方便使用、统一更改，先在 res\values\strings.xml 中定义本例中用到的新闻标题和新闻内容等字符串资源。参考代码如下：

```
<?xml version = "1.0" encoding = "utf-8"?>
<resources>
<string name = "app_name">01_TestTextView</string>
<string name = "title">培训机构增语文课减英语课</string>
<string name = "content">\t英语分值大幅缩水…学校都表示英语培训班的授课内容将向听力进一步倾斜,班级数量有可能会缩减。</string>
<string name = "link">详情</string>
</resources>
```

(3) 定义样式

在一个应用程序中为了使得文本格式统一，可以在"res\values\styles.xml"中预先定义好样式。本例使用的样式参考代码如下：

```
<resources xmlns:android = "http://schemas.android.com/apk/res/android">
<!-- SDK4.2自动生成的Application theme. -->
<style name = "AppTheme" parent = "AppBaseTheme">
</style>
    <!-- 自定义标题的样式 -->
    <style name = "title_style">
    <item name = "android:textColor">#00f</item>
    <item name = "android:textSize">20sp</item>
    <item name = "android:textStyle">bold</item>
    </style>
    <!-- 自定义内容的样式 -->
    <style name = "content_style">
    <item name = "android:textColor">#000</item>
    <item name = "android:textSize">15sp</item>
    </style>
</resources>
```

(4) 定义布局

从图 3-1a 中可看出，在本例的首页面中共有 3 个文本，从上到下分别用来显示新闻标题、新闻内容和新闻详情的链接，所以在 res\layout\activity_main.xml 中，使用垂直方向的线性布局管理这 3 个组件即可。

可参考表 3-1 设置布局及 TextView 组件的属性。

表 3-1 实例 1 中各组件的属性列表

序号	组件类型	属性名称	属性值	说明
1	LinearLayout	android:layout_width	match_parent	宽度匹配父容器
		android:layout_height	match_parent	高度匹配父容器
		android:orientation	vertical	垂直方向
2	TextView（新闻标题）	android:layout_width	match_parent	宽度匹配父容器
		android:layout_height	wrap_content	高度适合内容即可
		android:text	@string/title	设置文本为"string/title"的内容
		android:id	@+id/tv_title	增加 id 为"tv_title"的组件
		style	@style/title_style	使用 style/title_style 样式
		android:gravity	center_horizontal	文本在组件中水平方向居中对齐
3	TextView（新闻内容）	android:layout_width	match_parent	宽度匹配父容器
		android:layout_height	wrap_content	高度适合内容即可
		android:text	@string/content	设置文本为"string/content"的内容
		android:id	@+id/tv_content	增加 id 为"tv_content"的组件
		style	@style/content_style	使用 style/content_style 样式
4	TextView（详情）	android:id	@+id/tv_link	增加 id 为"tv_link"的组件
		android:layout_width	wrap_content	宽度适合内容即可
		android:layout_height	wrap_content	高度适合内容即可
		android:text	@string/link	设置文本为"string/link"的内容
		android:layout_gravity	right	将文本显示框与父容器右对齐

参考代码如下：

```xml
<LinearLayout xmlns:android = "http://schemas.android.com/apk/res/android"
    xmlns:tools = "http://schemas.android.com/tools"
    android:layout_width = "match_parent"
    android:layout_height = "match_parent"
    android:orientation = "vertical" >
    <!-- 使用 title_style 样式 -->
    <TextView
        android:gravity = "center_horizontal"
        android:layout_width = "match_parent"
        android:layout_height = "wrap_content"
        android:text = "@string/title"
        android:id = "@+id/tv_title"
        style = "@style/title_style" />
    <!-- 使用 content_style 样式 -->
    <TextView
        android:layout_width = "match_parent"
        android:layout_height = "wrap_content"
        android:text = "@string/content"
        android:id = "@+id/tv_content"
        style = "@style/content_style" />
    <!-- 右对齐链接文本 -->
    <TextView
        android:id = "@+id/tv_link"
```

```
        android:layout_width = "wrap_content"
        android:layout_height = "wrap_content"
        android:text = "@string/link"
        android:layout_gravity = "right"/>
</LinearLayout>
```

（5）在 Activity 中实现 TextView 文本的超链接

如前文所述，如果在 TextView 中有显式的链接文本，如 "http:\\www.baidu.com"，通过设置 android:autoLink = "all" 即可完成链接，但本例中链接文本是 "详情"，具体的链接目标需要在 Activity 中使用 SpannableString 对象来设置。

参考代码如下：

```
public class MainActivity extends Activity {
    TextView tv;
    @Override
    protected void onCreate(BundlesavedInstanceState) {
    super.onCreate(savedInstanceState);
        setContentView(R.layout.activity_main);
        //取得 id 为 tv_link 的文本显示框
        tv = (TextView)this.findViewById(R.id.tv_link);
        //获取 tv_link 中的文本
        String text = tv.getText().toString();
        //用 tv_link 中的文本值生成一个 SpannableString 对象
        SpannableString sp = new SpannableString(text);
        //设置超链接
        sp.setSpan(new URLSpan("http://beijing.qianlong.com" +
            "/3825/2013/10/23/7044@9071383.htm"),
            0,sp.length(),Spanned.SPAN_EXCLUSIVE_EXCLUSIVE);
        //SpannableString 对象设置给 TextView
        tv.setText(sp);
        //设置 TextView 可跳转到相关页面
        tv.setMovementMethod(LinkMovementMethod.getInstance());
    }
}
```

SpannableString 类可以用来定义 "内容不变，但标记对象可以附加和分离的文本对象"。常用方法有：

```
setSpan(Object what,int start,int end,int flags)
```

参数意义：

第 1 个参数用来设置标记对象链接目标或显示效果，第 2 个参数和第 3 个参数分别用来设置标记对象的起止位置（位置从 0 开始），第 4 个参数用来设置在标记对象前后输入新的字符时是否也应用此效果。

例如，以下代码可在点击标记对象时拨打电话 "4008123"：

```
sp.setSpan(newURLSpan("tel:4008123"),0,3,Spanned.SPAN_EXCLUSIVE_EXCLUSIVE);
```

再如，以下代码即可设置字体大小为默认值的 0.5 倍：

sp.setSpan(newRelativeSizeSpan(0.5f),8,10,Spanned.SPAN_EXCLUSIVE_EXCLUSIVE);

3.1.2 文本显示组件（TextView）的定义与使用

1. TextView 的定义

TextView 提供了一个类似标签的显示操作，主要功能在于显示文本，所以也被开发人员称为文本显示框。此类定义如下：

```
java.lang.Object
    ↳ android.view.View
        ↳ android.widget.TextView
```

在使用这个组件时，需要首先在 res/layout 中的 XML 文件里进行注册，注册时可参考如下代码：

```
<TextView
    android:layout_width = "wrap_content"
    android:layout_height = "wrap_content"
    android:text = "@string/hello_world" />
```

2. TextView 的常用属性

与布局管理器相似，TextView 的常用属性中也包括 android:layout_width 和 android:layout_height，分别表示 TextView 的宽度和高度。除此之外，它还有一些与文本显示有关的常用属性，如表 3-2 所示。

表 3-2　TextView 的常用属性说明

序号	属性名称	说明	编码举例
1	android:text	设置要显示的文本	将文本设置为 strings.xml 中名为"hello_world"的字符串： android:text = "@string/hello_world"
2	android:textSize	设置文本字号大小	将文本大小设置为"30sp"： android:textSize = "30sp"
3	android:textColor	设置文本颜色	将文本颜色设置为"绿色"： android:textColor = "#0f0"
4	android:textStyle	设置文本样式	文本使用"加粗"样式： android:textStyle = "bold"
5	android:autoLink	设置当文本为 Url 链接或电话号码或电子邮件地址时，文本显示为可点击的链接 可选值：none/web/email/phone/map/all	将文本中的电话设为可点击的链接： android:autoLink = "phone" 将文本中的电话、电子邮件地址、网址等均作为可点击的链接： android:autoLink = "all"

3.2 实例 2：微信登录

3.2.1 功能要求与操作步骤

1. 功能要求

完成如图 3-2 所示的应用程序。要求能够：

① 在图 3-2a 所示的界面中，两个编辑框分别提示用户输入 "QQ 号/微信号/手机号" 和 "密码" 的信息。用户输入信息的效果如图 3-2b 所示。

② 单击 "忘记密码"，可以打开相关链接，如图 3-2c 所示。

③ 当用户在第 1 个编辑框中输入的信息为 "12345678" 且密码为 "8765" 时，单击 "登录"，提示登录成功，效果如图 3-2d 所示；否则提示 "对不起，您的输入有误，请重试！"，并由第 1 个编辑框获取焦点，效果如图 3-2e 所示。

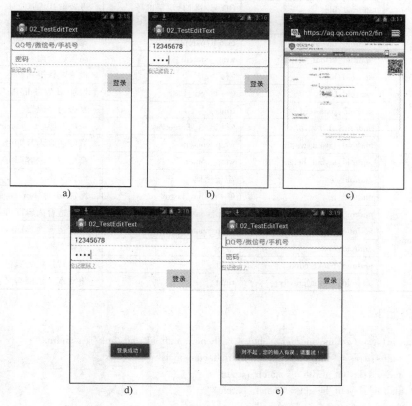

图 3-2 "登录微信" 运行效果图

2. 操作步骤

（1）创建项目

参考如下输入信息创建 Android 项目：

Application Name：02_TestEditText
Project Name：02_TestEditText

```
Package Name：com. book. testedittext
Activity Name：MainActivity
Layout Name：activity_main
```

(2) 定义布局

从图 3-2 中可看出，本应用可以使用垂直线性布局或者相对布局来实现，此处仅以垂直线性布局为例来定义布局和组件。可参考表 3-3 设置各个组件的属性。

表 3-3 实例 2 中各组件的属性列表

序号	组件类型	属性名称	属性值	说明
1	LinearLayout	android：layout_width	match_parent	宽度匹配父容器
		android：layout_height	match_parent	高度匹配父容器
		android：orientation	vertical	垂直方向
2	EditText	android：layout_width	match_parent	宽度匹配父容器
		android：layout_height	wrap_content	高度适合内容即可
		android：hint	QQ 号/微信号/手机号	设置默认提示文本
		android：id	@+id/edt_ID	增加 id 为"edt_ID"的组件
3	EditText	android：layout_width	match_parent	宽度匹配父容器
		android：layout_height	wrap_content	高度适合内容即可
		android：hint	密码	设置默认提示文本
		android：id	@+id/edt_Pss	增加 id 为"edt_Pss"的组件
		android：password	true	是否为密码框
4	TextView	android：id	@+id/tv_ForgetPss	增加 id 为"tv_ForgetPss"的组件
		android：layout_width	wrap_content	宽度适合内容即可
		android：layout_height	wrap_content	高度适合内容即可
		android：text	忘记密码	设置文本
5	Button	android：id	@+id/btn_Login	增加 id 为"btn_Login"的组件
		android：layout_width	wrap_content	宽度适合内容即可
		android：layout_height	wrap_content	高度适合内容即可
		android：text	登录	设置文本
		android：background	#0f0	设置背景色为绿色
		android：layout_gravity	right	将文本显示框与父容器右对齐

参考代码如下：

```
< LinearLayout xmlns：android = "http：//schemas. android. com/apk/res/android"
    xmlns：tools = "http：//schemas. android. com/tools"
    android：layout_width = "match_parent"
    android：layout_height = "match_parent"
    android：orientation = "vertical"
>

    < EditText
        android：id = "@+id/edt_ID"
        android：layout_width = "match_parent"
        android：layout_height = "wrap_content"
        android：hint = "QQ 号/微信号/手机号" >
```

```xml
<!--此编辑框获得焦点 -->
<requestFocus />
</EditText>
<EditText
        android:id = "@+id/edt_Pss"
        android:layout_width = "match_parent"
        android:layout_height = "wrap_content"
        android:hint = "密码"
        android:password = "true" >
</EditText>
<TextView
        android:id = "@+id/tv_ForgetPss"
        android:layout_width = "wrap_content"
        android:layout_height = "wrap_content"
        android:text = "忘记密码?"
        />
<Button
        android:id = "@+id/btn_Login"
        android:layout_width = "wrap_content"
        android:layout_height = "wrap_content"
        android:text = "登录"
        android:background = "#0f0"
        android:layout_gravity = "right" />
</LinearLayout>
```

（3）实现 MainActivity 中的功能

在本应用中，单击"忘记密码"时链接到相关网站的方法与第3.1节完全相同。在单击"登录"按钮时，需要根据用户输入信息来决定给出提示内容。为了使按钮可以响应单击事件，需要为按钮添加单击监听。添加监听的方法是：在Java文件中，调用并实现"setOnClickListener(View. OnClickListener 1)"。

参考代码如下：

```java
public class MainActivity extends Activity {
    //声明组件变量
    EditText edtID,edtPss;
    TextView txtForgetPss;
    Button btnLogin;
    //定义正确的用户信息
    String id = "12345678" , pss = "8765";
    @Override
    protected void onCreate( BundlesavedInstanceState) {
        super.onCreate( savedInstanceState) ;
        setContentView( R. layout. activity_main) ;
        //获取各个组件
        edtID = ( EditText)this. findViewById( R. id. edt_ID) ;
        edtPss = ( EditText)this. findViewById( R. id. edt_Pss) ;
        txtForgetPss = ( TextView)this. findViewById( R. id. tv_ForgetPss) ;
        btnLogin = ( Button)this. findViewById( R. id. btn_Login) ;
        //实现超链接
```

```
            String text = txtForgetPss.getText().toString();
            SpannableString sp = new SpannableString(text);
            //设置超链接
            sp.setSpan(new URLSpan("https://aq.qq.com/cn2/findpsw/pc/pc_find_pwd_input_a
                    ccount?pw_type=0&aquin="),0,sp.length(),
                    Spanned.SPAN_EXCLUSIVE_EXCLUSIVE);
            //SpannableString 对象设置给 TextView
            txtForgetPss.setText(sp);
            //设置 TextView 可跳转到相关页面
            txtForgetPss.setMovementMethod(LinkMovementMethod.getInstance());
            //为按钮设置单击监听,并在 onClick 中实现
            btnLogin.setOnClickListener(new OnClickListener(){
                @Override
                public void onClick(View v){
                    // TODO Auto-generated method stub
                    if(edtID.getText().toString().equals(id)&&
                        edtPss.getText().toString().equals(pss)){
                        //给出"登录成功"的提示
                        Toast.makeText(MainActivity.this,"登录成功!",
                                    Toast.LENGTH_SHORT).show();
                    }else{
                    //给出错误提示
                    Toast.makeText(MainActivity.this,"对不起,您的输入有误,请重试!",
                            Toast.LENGTH_SHORT).show();
                        //清空编辑框文本
                    edtID.setText(null);
                    edtPss.setText(null);
                    //获取焦点
                    edtID.requestFocus();
                    }
                }
            });
        }
    }
```

3.2.2 编辑框(EditText)的定义与使用

1. EditText 的定义

如果应用中需要添加用来输入文本或编辑文本的组件,以便达到很好的人机交互和用户体验,那么这个组件非编辑框 EditText 莫属。此类定义如下:

```
java.lang.Object
   ↳ android.view.View
       ↳ android.widget.TextView
           ↳ android.widget.EditText
```

在使用这个组件时,需要首先在 res/layout 中的 XML 文件里进行注册,注册时可参考如下代码:

```
<EditText
    android:id="@+id/edtTxtName"
```

```
android:layout_width = "wrap_content"
android:layout_height = "wrap_content" >
```

2. EditText 的常用方法

从类的定义中可以发现，EditText 继承于 TextView，所以二者具有很多相同的属性，对于文本的各种操作也可以在此类中继续使用。EditText 的常用方法如表 3-4 所示。

表 3-4 EditText 的常用方法说明

序号	方法名称	说明
1	public EditablegetText( )	获取编辑框中的文本
2	public final voidsetText(CharSequence text)	继承自 TextView 类的方法，用于设置编辑框中的文本为 text
3	public final voidrequestFocus( )	继承自 View 类的方法，用于为编辑框获取焦点，当取得焦点成功时返回 true，否则返回 false

3.2.3 按钮（Button）的定义与使用

1. Button 的定义

按钮是在人机交互界面上使用最多的组件，当提示用户进行某些选择时，就可以通过相应的按钮操作来接收用户的选择。在 Android 中，使用 Button 可以定义出一个按钮，并且可以在按钮上指定相应的显示文字，此类定义如下：

```
java.lang.Object
    ↳ android.view.View
        ↳ android.widget.TextView
            ↳ android.widget.Button
```

通过此类的定义可以发现，Button 是 TextView 的子类，实际上所谓的按钮就是一个特殊的文本组件，此类中定义的属性与 TextView 相同。

在使用这个组件时，需要首先在 res/layout 当中的 XML 文件里进行注册，注册时可参考如下代码：

```
< Button
    android:id = "@+id/btn_Login"
    android:layout_width = "wrap_content"
    android:layout_height = "wrap_content"
    android:text = "登录"
    android:onClick = "clickLogin" />
```

需要说明的是，在上述注册代码中的 onClick 属性不是必须要注册的，但是一经注册，必须在 Java 文件中实现，否则会在单击按钮时出现"java.lang.NoSuchMethodException"的异常。在 Java 文件中实现时要求方法的可见性为"public"，而且方法有且只有一个 View 类型的参数，参考代码如下：

```
publicvoid clickLogin( View v) {
    //具体功能逻辑
    ...
}
```

如果在 XML 中没有对 onClick 属性进行注册声明，为了使用户在单击按钮时能够有所响应，可以采用传统的添加监听的方法，参考代码如下：

```
final Button button = (Button)findViewById(R.id.btn_Login);
button.setOnClickListener(new View.OnClickListener(){
    public void onClick(View v){
        //具体功能逻辑
    }
});
```

2. Button 的常用方法

Button 的常用方法如表 3-5 所示。

表 3-5　Button 的常用方法说明

序号	方法名称	说　明
1	public void setOnClickListener(View.OnClickListener l)	继承于 View 类的方法，为单击事件注册监听
2	public abstract void onClick(View v)	在此方法中实现单击时的具体处理工作

3.2.4　信息提示框（Toast）使用简介

1. Toast 的定义

信息提示框 Toast 是 Android 中的一个以简单提示信息为主要显示操作的组件，它没有焦点，在提示用户时不会打断用户的正常操作，而且显示的时间有限，经过一定的时间后就会自动消失。此类定义如下：

```
java.lang.Object
  ↳ android.widget.Toast
```

2. Toast 的常用方法及常量

在显示信息提示框时，需要使用到一些方法和常量，如表 3-6 所示。

表 3-6　Toast 的常用方法说明

序号	方法及常量	说　明
1	public static final int LENGTH_LONG	显示时间较长
2	public static final int LENGTH_SHORT	显示时间较短
3	public static Toast makeText(Context context, CharSequence text, int duration)	创建一个 Toast 对象，其中第 1 个参数指定了应用程序的环境，第 2 个参数指定了显示的文本为 text，第 3 个参数指定了 duration 为显示时间
4	public void show()	显示 Toast 对象

3.3　实例 3：注册应用账号

3.3.1　功能要求与操作步骤

1. 功能要求

完成如图 3-3 所示的"注册应用账号"应用程序。要求能够：

① 应用程序初次启动，显示如图 3-3a 所示的界面。

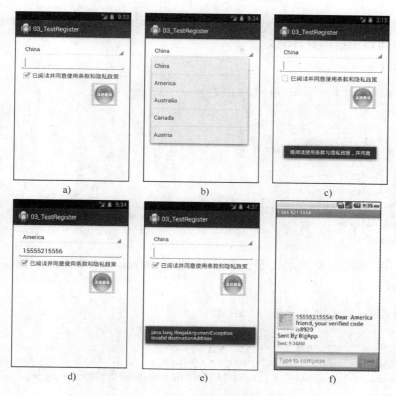

图3-3 "注册应用账号"运行效果图

② 在图3-3a所示的页面中,使用到的组件依次是下拉列表框,输入电话号码的文本编辑框和复选框(默认为被选中状态)。

③ 当单击下拉列表框时,页面效果如图3-3b所示。

④ 单击右下角的图形按钮"注册登录"时,如果复选框未勾选,则弹出如图3-3c所示信息框;否则向输入的电话号码发送验证码等信息。说明:实际的应用中,验证码应由服务器向指定的电话号码发送,此处运行本应用的模拟器将模拟执行此过程,模拟器电话格式如图3-3d所示。

⑤ 当目标电话号码格式不正确或为空时,提示错误,如图3-3e所示。

⑥ 启动第2个模拟器,用于接收发送的信息,收到短信的效果如图3-3f所示。

2. 操作步骤

(1)创建项目

参考如下输入信息创建Android项目:

> Application Name:03_TestRegister
> Project Name:03_TestRegister
> Package Name:com. book. testregister
> Activity Name:MainActivity
> Layout Name:activity_main

(2)准备字符串资源和图片资源

本例中下拉表需要用到多个字符串资源,可以预先将其存储在字符串数组中。在res/

values/ 目录下，新建 country_data.xml 文件。在文件中，定义名为"country_name"的数组资源，数组中元素可以根据需要自定义。

参考代码如下：

```xml
<?xml version="1.0" encoding="utf-8"?>
<resources>
<string-array name="country_name">
<item>China</item>
<item>America</item>
<item>Australia</item>
<item>Canada</item>
<item>Austria</item>
</string-array>
</resources>
```

除此之外，复选框需要提示文字，在 res/values/strings.xml 中添加相应字符串资源即可，参考代码如下：

```xml
<?xml version="1.0" encoding="utf-8"?>
<resources>
<string name="app_name">03_TestRegister</string>
<string name="action_settings">Settings</string>
<string name="agreePolicy">已阅读并同意使用条款和隐私政策</string>
</resources>
```

本例中，图片按钮所需的图片资源名称为"register.gif"，将其保存在 res/drawable-x 中，以备使用。

（3）定义布局

除布局资源管理器外，本应用还需要用到 4 个组件，分别是 Spinner、EditText、CheckBox 和 ImageBox。各个组件的属性可以参考表 3-7 设置。

表 3-7 实例 3 中各组件的属性列表

序号	组件类型	属性名称	属性值	说明
1	RelativeLayout	android:layout_width	match_parent	宽度匹配父容器
		android:layout_height	match_parent	高度匹配父容器
2	Spinner	android:id	@+id/spn_country	增加 id 为"spn_country"的组件
		android:layout_height	wrap_content	高度适合内容即可
		android:layout_width	match_parent	宽度匹配父容器
		android:layout_alignParentTop	true	与父容器顶部对齐
		android:entries	@array/country_name	设置使用的文本资源来自于数组 country_name
3	EditText	android:id	@+id/edt_phone	增加 id 为"edt_phone"的组件
		android:layout_height	wrap_content	高度适合内容即可
		android:layout_width	match_parent	宽度匹配父容器
		android:layout_below	@+id/spn_country	在指定组件的下方
		android:layout_alignLeft	@+id/spn_country	与指定组件左对齐
		android:inputType	phone	设置输入类型为电话

（续）

序号	组件类型	属性名称	属性值	说明
4	CheckBox	android:id	@+id/chk_agreePolicy	增加 id 为"chk_agreePolicy"的组件
		android:layout_width	wrap_content	宽度适合内容即可
		android:layout_height	wrap_content	高度适合内容即可
		android:text	@string/agreePolicy	设置文本
		android:layout_alignLeft	@+id/edt_phone	与指定组件左对齐
		android:layout_below	@+id/edt_phone	在指定组件下方
		android:checked	true	默认为被选中
5	ImageButton	android:id	@+id/imgbtn_register	增加 id 为"imgbtn_register"的组件
		android:layout_width	wrap_content	宽度与内容匹配即可
		android:layout_height	wrap_content	高度与内容匹配即可
		android:layout_alignRight	@+id/chk_agreePolicy	与指定组件右对齐
		android:layout_below	@+id/chk_agreePolicy	在指定组件的下方
		android:src	@drawable/register	设置图片
		android:onClick	clickReg	添加单击监听的方法为"clickReg"

参考代码：

```
<RelativeLayout xmlns:android = "http://schemas.android.com/apk/res/android"
    xmlns:tools = "http://schemas.android.com/tools"
    android:layout_width = "match_parent"
    android:layout_height = "match_parent"
    >
    <Spinner
        android:id = "@+id/spn_country"
        android:layout_width = "match_parent"
        android:layout_height = "wrap_content"
        android:layout_alignParentTop = "true"
        android:entries = "@array/country_name"
        />
    <EditText
        android:id = "@+id/edt_phone"
        android:layout_width = "match_parent"
        android:layout_height = "wrap_content"
        android:layout_below = "@+id/spn_country"
        android:layout_alignLeft = "@+id/spn_country"
        android:inputType = "phone" >
    </EditText>
    <CheckBox
        android:id = "@+id/chk_agreePolicy"
        android:layout_width = "wrap_content"
        android:layout_height = "wrap_content"
        android:layout_alignLeft = "@+id/edt_phone"
        android:layout_below = "@+id/edt_phone"
        android:text = "@string/agreePolicy"
        android:checked = "true" />
    <ImageButton
```

```
                    android:id = "@+id/imgbtn_register"
                    android:layout_width = "wrap_content"
                    android:layout_height = "wrap_content"
                    android:layout_alignRight = "@+id/chk_agreePolicy"
                    android:layout_below = "@+id/chk_agreePolicy"
                    android:src = "@drawable/register"
                    android:onClick = "clickReg"/>
            </RelativeLayout>
```

需要说明的是,第一,如果希望编辑框中输入符合特定格式的信息,可以通过设置 EditText 的 android:inputType 属性来实现,其值可以是 textEmailAddress(电子邮箱格式)、textPersonName(姓名格式)、textPassword(密码格式)、number(数字格式)、datetime(日期时间格式)等。例如上述代码中:

```
                    android:inputType = "phone"
```

使得本编辑框中只能输入符合电话号码格式的信息。

第二,如果在 XML 文件中为按钮设置了 android:onClick 属性,一定需要在 Java 文件中实现以该属性值为名称的方法,具体格式前文已述,此处不再重复。

(4)实现 MainActivity 中的功能

在本应用中,为了记录用户在 Spinner 下拉列表中的选项,需要为 Spinner 添加 setOnItemSelectedListener 监听。方便起见,MainActivity 类直接实现了 OnItemSelectedListener 接口。

注册过程中需要向目标电话发送随机生成的 4 位验证码。随机生成 4 位验证码的原理为:分别生成 4 个 10 以内的数字(0~9,不含 10),然后将这 4 个随机数在字符串中拼接成为验证码。

参考代码如下:

```java
public class MainActivity extends Activity implements OnItemSelectedListener {
    Spinner spnCountry;
    EditText edtPhone;
    CheckBox chkAgreePolicy;
    //countryName 用来记录用户所选的国家名称
    //phone 用来记录用户输入的电话号码
    //content 用来拼接短信字符串
    String countryName = null, phone = null, content = null;
    @Override
    protected void onCreate(Bundle savedInstanceState) {
        super.onCreate(savedInstanceState);
        setContentView(R.layout.activity_main);
        spnCountry = (Spinner)this.findViewById(R.id.spn_country);
        edtPhone = (EditText)this.findViewById(R.id.edt_phone);
        chkAgreePolicy = (CheckBox)this.findViewById(R.id.chk_agreePolicy);
        //为下拉列表添加选择监听
        spnCountry.setOnItemSelectedListener(this);
    }
    public void clickReg(View v) {
```

```java
            //如果未勾选同意政策,则给用户以操作提示,否则发送信息
            if(!chkAgreePolicy.isChecked()){
                Toast.makeText(MainActivity.this,
                        "请阅读使用条款与隐私政策,并同意",
                        Toast.LENGTH_LONG).show();
            }else{
                //捕获电话号码格式不正确的异常
                try{
                    //取得目标电话号码,为发送短信做准备
                    phone = edtPhone.getText().toString();
                    //生成四位随机数,做验证码
                    Random r = new Random();
                    r.setSeed(System.currentTimeMillis());
                    int verifyTemp = 0;
                    String verifyCode = "";
                    for(int i=0;i<4;i++){
                        verifyTemp = r.nextInt(10);
                        verifyCode += verifyTemp;
                    }
                    content = "Dear  " + countryName + " friend," +
                            " your verified code is" + verifyCode + " \nSent By BigApp";
                    //发送短信
                    SmsManager smsManager = SmsManager.getDefault();
                    smsManager.sendTextMessage(
                            phone,
                            null,
                            content,
                            null,
                            null);
                }catch(Exception e){
                    // 在信息提示框中输出异常描述
                    Toast.makeText(MainActivity.this,
                            String.valueOf(e),
                            Toast.LENGTH_LONG).show();
                }
            }
        }
    }
    @Override
    public void onItemSelected(AdapterView<?> parent,View view,int position,
            long itemID){
        //当下拉列表框中某项被选中时,将该项的值写到 countryName 字符串中
        countryName = parent.getItemAtPosition(position).toString();
    }
    @Override
    public void onNothingSelected(AdapterView<?> arg0){
        // TODO Auto-generated method stub
    }
}
```

(5) 为发送短信添加权限

为了使应用程序可以正常发送短信,需要在 AndroidManifest.xml 文件中添加 "android.

permission. SEND_SMS"的使用权限。

参考代码如下：

```
<uses-permission android:name="android.permission.SEND_SMS"/>
```

配置完成后，启动两个模拟器，通常其端口号分别为5554和5556。在测试本应用时，使用其中一个模拟器向另外一个模拟器发送短信，便于查看运行效果。

3.3.2 下拉列表框（Spinner）的定义与使用

1. Spinner 的定义

下拉列表框 Spinner 也是一种常见的图形组件，它可以为用户提供列表的选择方式，比较节省手机的屏幕空间。此类定义如下：

```
java.lang.Object
  ↳ android.view.View
    ↳ android.view.ViewGroup
      ↳ android.widget.AdapterView<T extends android.widget.Adapter>
        ↳ android.widget.AbsSpinner
          ↳ android.widget.Spinner
```

在使用这个组件时，需要首先在 res/layout 中的 XML 文件里进行注册，注册时可参考如下代码：

```
<Spinner
    android:id="@+id/spn_country"
    android:layout_width="match_parent"
    android:layout_height="wrap_content"
    android:entries="@array/country_name"
/>
```

与其他组件相同，需要为 Spinner 声明其宽度、高度和 id，除此之外，还需要额外定义其数据项来源，用到了 android:entries 属性，其值为数组资源。

2. Spinner 的常用方法

Spinner 常用的方法大都继承自 AdapterView，具体如表3-8所示。

表3-8 Spinner 的常用方法说明

序号	方法名称	说明
1	public void setOnItemSelectedListener(AdapterView.OnItemSelectedListener listener)	继承于 AdapterView 的方法，为视图组件注册选择监听
2	public abstract void onItemSelected(AdapterView<?> parent, View view, int position, long id)	在此方法中实现某数据项被选中时的具体处理工作，常用到第1个参数和第3个参数。其中，第1个参数 parent 用来表明选择了哪个视图组件，第3个参数表明选择了视图组件哪个位置的数据项（位置从0开始）。在实现时，可以调用 getItemAtPosition(position) 方法取得与选择项有关的数据信息
3	public Object getItemAtPosition(int position)	继承于 AdapterView 的方法，参数为选择项的位置，返回值为与选择项有关的数据信息

在 Java 文件中也可以使用 ArrayAdapter 类完成 Spinner 的数据项添加工作。ArrayAdapter 主要有两个功能：读取资源文件中定义的列表项和通过 List 集合设置列表项。例如，在 Java 文件中为颜色列表 spnColor 添加资源文件 arrays.xml 中名为"colors"的数组的数据项，并显示如图 3-4 所示的列表风格。

res/layout 中注册 Spinner 的参考代码如下：

```
<Spinner android:id = "@ + id/spn_color"
    android:layout_width = "match_parent"
    android:layout_height = "wrap_content"
/>
```

图 3-4 单选钮列表风格示意图

res/values/arrays.xml 中定义 colors 数组的参考代码如下：

```
<?xml version = "1.0" encoding = "utf-8"?>
<resources>
<string-array name = "colors">
<item>红色</item>
<item>绿色</item>
<item>灰色</item>
<item>橙色</item>
</string-array>
</resources>
```

Java 文件中的参考代码如下：

```
SpinnerspnColor = (Spinner)findViewById(R.id.spn_color);
ArrayAdapter<CharSequence> aaColor = ArrayAdapter.createFromResource(
            this,
            R.array.colors,
            android.R.layout.simple_list_item_1);
aaColor.setDropDownViewResource(android.R.layout.simple_list_item_single_choice);
spnColor.setAdapter(aaColor);
```

在上述代码中，"aaColor"是调用 ArrayAdapter 类的 createFromResource 方法而生成。在使用这个方法时涉及的参数依次是当前环境对象，列表中显示的数据项和列表的显示风格，其中显示风格可以自定义。

通过调用 ArrayAdapter 的 setDropDownViewResource 方法，还可以设置下拉列表项的显示风格，如此处使用了"android.R.layout.simple_list_item_single_choice"单选按钮风格。

3.3.3 复选框（CheckBox）的定义与使用

1. CheckBox 的定义

复选框是具有两种选择状态的特殊按钮组件，其主要功能是帮助用户完成复选操作。在用户输入信息时，可以一次选择多个内容，并且可以随时更改其选择状态。此类定义如下：

```
java.lang.Object
    ↳ android.view.View
```

↳android.widget.TextView
↳android.widget.Button
↳android.widget.CompoundButton
↳android.widget.CheckBox

在使用这个组件时，需要首先在 res/layout 中的 XML 文件里进行注册，注册时可参考如下代码：

```
<CheckBox
    android:id = "@+id/chk_agreePolicy"
    android:layout_width = "wrap_content"
    android:layout_height = "wrap_content"
    android:text = "@string/agreePolicy"
    android:checked = "true"/>
```

在 xml 文件中注册时，可以使用 android:text 属性设置复选框的提示文本，还可以通过设置 android：checked 的值决定其默认的选择状态。

2. CheckBox 的常用方法

CheckBox 常用的方法大都继承自 CompoundButton，具体如表 3-9 所示。

表 3-9 CheckBox 的常用方法说明

序号	方法名称	说明
1	public boolean isChecked( )	继承于 CompoundButton 的方法，用于查看复选框是否被勾选。勾选时，返回值为 true；未勾选时返回 false
2	public void setChecked(boolean checked)	继承于 CompoundButton 的方法，用于设置复选框的选择状态，参数为 true 或 false
3	public void setOnCheckedChangeListener(CompoundButton.OnCheckedChangeListener listener)	继承于 CompoundButton 的方法，用于为复选框添加选择状态改变的监听
4	public abstract void onCheckedChanged(CompoundButton buttonView, boolean isChecked)	在此方法中实现复选框状态发生改变时的具体处理工作。第 1 个参数用来指明哪个复选框的状态发生了改变，第 2 个参数用来表明改变后的状态值

3.3.4 图像按钮（ImageButton）的定义与使用

1. ImageButton 的定义

与 Button 相似，在 Android 中还提供了一个图片按钮组件 ImageButton。此类定义如下：

```
java.lang.Object
↳android.view.View
    ↳android.widget.ImageView
        ↳android.widget.ImageButton
```

在使用这个组件时，首先在 res/layout 中的 XML 文件里进行注册，注册时可参考如下代码：

```
<ImageButton
    android:id = "@+id/imgbtn_register"
```

```
android:layout_width = "wrap_content"
android:layout_height = "wrap_content"
android:src = "@drawable/register"
android:onClick = "clickReg"/>
```

在 xml 文件中，可以使用 android:src 属性设置 ImageButton 所显示的图片资源，为了使用方便，需要将准备好的图片存放在 res/drawable – x 文件夹中。与 Button 相似，ImageButton 也可以通过设置 android:onClick 来注册单击事件发生时执行的方法。

2. ImageButton 的常用方法

ImageButton 常用的方法如表 3-10 所示。

表 3-10　< ImageButton > 的常用方法说明

序号	方法名称	说　　明
1	public void setImageResource(int resId)	继承于 ImageView 的方法，用于为图片按钮设置图片资源，参数为图片资源 ID
2	public void setOnClickListener(View. OnClickListener l)	继承于 View 类的方法，为单击事件注册监听
3	public abstract void onClick(View v)	在此方法中实现单击时的具体处理工作

3.3.5　短信管理器（SmsManager）使用简介

1. SmsManager 的定义

使用短信管理器 SmsManager 类可以实现短信的管理功能，最常用的就是通过此类发送短信。此类定义如下：

```
java. lang. Object
    ↳ android. telephony. SmsManager
```

2. SmsManager 的常用方法

SmsManager 常用的方法如表 3-11 所示。

表 3-11　SmsManager 的常用方法说明

序号	方法名称	说　　明
1	public static SmsManager getDefault( )	取得默认的短信管理器实例
2	public ArrayList < String > divideMessage(String text)	将信息文本 text 分成若干片段，以使得信息长度小于短信所能包含字符的最大值，注意：text 不可为空，否则将抛出异常，返回数组列表
3	public void sendTextMessage (String destinationAddress, String scAddress, Stringtext, PendingIntent sentIntent, PendingIntent deliveryIntent)	使用 SMS 发送文本信息。第 1 个参数是接收接收方地址；第 2 个参数是服务器地址，如果为空则表示当前设备；第 3 个参数是发送的信息；第 4 个参数如果不为空，表示当消息发出时，通过 PendingIntent 来广播发送成功或者失败的信息报告；第 5 个参数如果不为空，表示当成功发送信息后 PendingIntent 要进行广播

3. 发送短信的基本步骤

在发送短信时，需要按照以下步骤进行：

（1）取得短信管理器

调用 SmsManager 类的 getDefault( ) 方法，可以取得默认的短信管理器，其返回值为 Sms-

Manager 类型。

参考代码如下：

```
SmsManager smsManager = SmsManager.getDefault();
```

（2）准备短信文本

短信文本是一个字符串，根据需要将需要发送的字符串连接，如果长度较大，可以调用 SmsManager 类的 divideMessage(String) 方法，对原始字符串进行切割分解。

例如，当短信长度大于 70 个字符时，拆分信息。参考代码如下：

```
if(content.length() > 70){
    List<String> msgs = smsManager.divideMessage(content);
    Iterator<String> iter = msgs.iterator();
    while(iter.hasNext()){
        String msg = iter.next();
        //将 msg 发送到接收方
    }
}
```

（3）发送短信文本

调用 SmsManager 类的 sendTextMessage(String destinationAddress, String scAddress, String text, PendingIntent sentIntent, PendingIntent deliveryIntent) 方法发送信息。

在发送短信的过程中，可能会由于电话号码有误或其他原因造成各种异常，为了良好的用户体验，可以将发送短信的过程封装在异常捕获模块中。参考代码如下：

```
try{
    smsManager.sendTextMessage(phone, null, content, null, null);
}catch(Exception e){
    // 在信息提示框中输出异常描述
    Toast.makeText(MainActivity.this, String.valueOf(e),
        Toast.LENGTH_LONG).show();
}
```

特别提醒，当允许应用程序发送短信时，需要在 AndroidManifest.xml 中添加 SMS 权限。

3.4 实例 4：完善个人资料

3.4.1 功能要求与操作步骤

1. 功能要求

完成如图 3-5 所示的应用程序。要求能够：

① 应用程序的主页面如图 3-5a 所示，性别的默认值是"男"。

② 用户可以在编辑框中分别输入用户名、密码及密码确认信息，还可以通过单击单选按钮和图像框选择性别和头像，如图 3-5b 所示。

③ 当用户输入的两次密码相同时，弹出如图 3-5c 所示的对话框，该对话框的图标为用

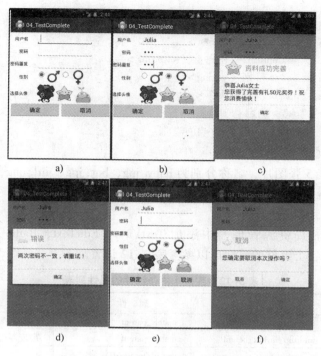

图 3-5 "完善个人资料"运行效果图

户所选的头像,并且在对话框中根据用户选择的性别显示"先生"或"女士"称呼,单击对话框中的"OK"按钮可以退出应用。

④ 当用户输入的两次密码不同时,弹出如图 3-5d 所示的对话框,单击对话框中的"确定"按钮后,即可清空"密码"和"密码重复"编辑框的文本,并且使"密码"编辑框获得焦点,效果如图 3-5e 所示。

⑤ 当用户在图 3-5a 所示的页面中单击"取消"按钮时,显示如图 3-5f 所示的对话框,单击对话框中的"取消"按钮则退回主页面,单击对话框中的"确定"按钮则退出应用。

2. 操作步骤

(1)创建项目

参考如下输入信息创建 Android 项目:

```
Application Name:04_TestComplete
Project Name:04_TestComplete
Package Name:com.book.testcomplete
Activity Name:MainActivity
Layout Name:activity_main
```

(2)准备字符串资源和图片资源

在 res/values/strings.xml 中定义经常用到的字符串资源,参考代码如下:

```
<?xml version="1.0" encoding="utf-8"?>
<resources>
<string name="app_name">04_TestComplete</string>
```

```
    <string name = "userID" >用户名</string>
    <string name = "pss" >密码</string>
    <string name = "pssConfirm" >密码重复</string>
    <string name = "sex" >性别</string>
    <string name = "image" >选择头像</string>
    <string name = "ok" >确定</string>
    <string name = "cancel" >取消</string>
</resources>
```

应用中使用到了5张图片，名称分别为baby.jpg、boy.jpg、girl.png、star.jpg、tree.jpg，这些图片都被放在res/drawable-x目录中。

（3）定义布局

在TableLayout布局资源管理器中添加TextView、EditText、RadioGroup、RadioButton、ImageView等组件。各个组件的属性可以参考表3-12设置。

表3-12 实例4中各组件的属性列表

序号	组件类型	属性名称	属性值	说明
1	TableLayout	android:layout_width	match_parent	宽度匹配父容器
		android:layout_height	match_parent	高度匹配父容器
2	TextView（各项提示信息）	android:id	@+id/tv_userID @+id/tv_pss @+id/tv_pssConfirm @+id/tv_sex @+id/tv_image	增加相应id的组件
		android:layout_height	wrap_content	高度适合内容即可
		android:layout_width	wrap_content	宽度适合内容即可
		android:gravity	right	与右侧的列边框对齐
		android:text	@string/userID @string/pss @string/pssConfirm @string/sex @string/image	设置文本
3	EditText（用户名编辑框）	android:id	@+id/edt_userID	增加id为"edt_phone"的组件
		android:layout_height	wrap_content	高度适合内容即可
		android:layout_width	wrap_content	宽度适合内容即可
		android:layout_marginLeft	28dp	左边距为28dp
4	EditText（密码和密码重复编辑框）	android:id	@+id/edt_pss @+id/edt_pssConfirm	增加相应id的组件
		android:layout_width	wrap_content	宽度适合内容即可
		android:layout_height	wrap_content	高度适合内容即可
		android:layout_marginLeft	28dp	左边距为28dp
		android:layout_inputType	textPassword	接收密码文本
5	RadioGroup	android:id	@+id/rdg_sex	增加id为"rdg_sex"的组件
		android:orientation	horizontal	水平方向排列子组件
		android:layout_marginLeft	28dp	左边距为28dp

(续)

序号	组件类型	属性名称	属性值	说明
6	RadioButton	android:id	@+id/rd_male @+id/rd_female	增加相应 id 的组件
		android:layout_width	wrap_content	宽度适合内容即可
		android:layout_heigh	wrap_content	高度适合内容即可
		android:drawableRight	@drawable/boy @drawable/girl	单选钮右侧放置相应图片资源
		android:checked	true	只为第1个单选钮设置此属性,表示默认性别为"男"
		android:gravity	left	左对齐
7	LinearLayout (放头像的子布局)	android:orientation	horizontal	水平方向排列子组件
		android:layout_width	wrap_content	宽度适合内容即可
		android:layout_height	wrap_content	高度适合内容即可
		android:layout_marginLeft	28dp	左边距为28dp
		android:gravity	center_horizontal	子组件在水平方向居中对齐
8	ImageView (头像框,上述 LinearLayout 的子组件)	android:id	@+id/img_baby @+id/img_star @+id/img_tree	增加相应 id 的组件
		android:layout_width	wrap_content	宽度适合内容即可
		android:layout_height	wrap_content	高度适合内容即可
		android:src	@drawable/baby @drawable/star @drawable/tree	设置各图像框的图片资源
		android:layout_margin	1dp	设置各图像框的边距
9	TableLayout (放置两个按钮的子布局)	android:layout_width	fill_parent	宽度匹配父容器
		android:layout_height	wrap_content	高度适合内容即可
		android:gravity	center_horizontal	子组件在水平方向居中对齐
		android:stretchColumns	*	平均分布各列
10	Button	android:id	@+id/btn_ok @+id/btn_cancel	增加相应 id 的组件
		android:layout_width	wrap_content	宽度适合内容即可
		android:layout_height	wrap_content	高度适合内容即可
		android:text	@string/ok @string/cancel	设置显示文本
		android:onClick	clickOK clickCancel	设置单击事件响应的方法

参考代码:

```
<TableLayout xmlns:android = "http://schemas.android.com/apk/res/android"
    android:layout_width = "match_parent"
    android:layout_height = "match_parent" >
    <TableRow>
        <TextView
            android:id = "@+id/tv_userID"
```

```xml
            android:layout_width = "wrap_content"
            android:layout_height = "wrap_content"
            android:text = "@string/userID"
            android:gravity = "right"/>
    <EditText
            android:id = "@+id/edt_userID"
            android:layout_width = "wrap_content"
            android:layout_height = "wrap_content"
            android:layout_marginLeft = "28dp"/>
</TableRow>
<TableRow>
    <TextView
            android:id = "@+id/tv_pss"
            android:layout_width = "wrap_content"
            android:layout_height = "wrap_content"
            android:text = "@string/pss"
            android:gravity = "right"/>
    <EditText
            android:id = "@+id/edt_pss"
            android:layout_width = "wrap_content"
            android:layout_height = "wrap_content"
            android:layout_marginLeft = "28dp"
            android:inputType = "textPassword" />
</TableRow>
<TableRow>
    <TextView
            android:id = "@+id/tv_pssConfirm"
            android:layout_width = "wrap_content"
            android:layout_height = "wrap_content"
            android:text = "@string/pssConfirm"
            android:gravity = "right"/>
    <EditText
            android:id = "@+id/edt_pssConfirm"
            android:layout_width = "wrap_content"
            android:layout_height = "wrap_content"
            android:layout_marginLeft = "28dp"
            android:inputType = "textPassword" />
</TableRow>
<TableRow>
    <TextView
            android:id = "@+id/tv_sex"
            android:layout_width = "wrap_content"
            android:layout_height = "wrap_content"
            android:text = "@string/sex"
            android:layout_gravity = "right|center_vertical"/>
    <RadioGroup
            android:id = "@+id/rdg_sex"
            android:orientation = "horizontal"
```

```xml
            android:layout_marginLeft = "28dp"  >
    <RadioButton
        android:id = "@+id/rd_male"
        android:layout_width = "wrap_content"
        android:layout_height = "wrap_content"
        android:drawableRight = "@drawable/boy"
        android:checked = "true"
        android:gravity = "left" />
    <RadioButton
        android:id = "@+id/rd_female"
        android:layout_width = "wrap_content"
        android:layout_height = "wrap_content"
        android:drawableRight = "@drawable/girl"
        android:gravity = "right"/>
    </RadioGroup>
</TableRow>
<TableRow>
<TextView
        android:id = "@+id/tv_image"
        android:layout_width = "wrap_content"
        android:layout_height = "wrap_content"
        android:text = "@string/image"
        android:layout_gravity = "right|center_vertical"/>
<LinearLayout
        android:orientation = "horizontal"
        android:layout_width = "wrap_content"
        android:layout_height = "wrap_content"
        android:layout_marginLeft = "28dp"
        android:gravity = "center_horizontal" >
<ImageView
        android:id = "@+id/img_baby"
        android:layout_width = "wrap_content"
        android:layout_height = "wrap_content"
        android:src = "@drawable/baby"
        android:layout_margin = "1dp"/>
    <ImageView
        android:id = "@+id/img_star"
        android:layout_width = "wrap_content"
        android:layout_height = "wrap_content"
        android:src = "@drawable/star"
        android:layout_margin = "1dp"/>
    <ImageView
        android:id = "@+id/img_tree"
        android:layout_width = "wrap_content"
        android:layout_height = "wrap_content"
        android:src = "@drawable/tree"
        android:layout_margin = "1dp"/>
</LinearLayout>
```

```xml
        </TableRow>
    <TableLayout
        android:layout_width = "fill_parent"
        android:layout_height = "wrap_content"
        android:gravity = "center_horizontal"
        android:stretchColumns = " * ">
    <TableRow >
        <Button
            android:id = "@+id/btn_ok"
            android:layout_width = "wrap_content"
            android:layout_height = "wrap_content"
            android:text = "@string/ok"
            android:onClick = "clickOK" />
        <Button
            android:id = "@+id/btn_cancel"
            android:layout_width = "wrap_content"
            android:layout_height = "wrap_content"
            android:text = "@string/cancel"
            android:onClick = "clickCancel"/>
    </TableRow >
    </TableLayout >
</TableLayout >
```

为了页面美观、整齐，实例 4 再次使用了嵌套布局，使用时需要注意各个布局管理器的嵌套关系，在 XML 文件中表现为各个布局管理器的起止标签的对应。

（4）实现 MainActivity 中的功能

在应用中，为了记录用户选择的单选按钮值，以及用户单击的图像框，需要为 RadioGroup 添加 setOnCheckedChangeListener 监听，为 ImageView 添加 setOnClickListener 监听。为了方便起见，"MainActivity" 类直接实现了 OnCheckedChangeListener 和 OnClickListener 接口。

需要注意的是，单击对话框中的各个按钮时，也可以响应具体处理，这就需要实现 OnClickListener 接口方法，但是必须在此接口前加 DialogInterface，以表明这个接口方法的来源。

参考代码如下：

```java
public class MainActivity extends Activity implements OnCheckedChangeListener,OnClickListener{
    EditText edtUserID,edtPssConfirm,edtPss;
    RadioGroup rdgSex;
    ImageView imgBaby,imgStar,imgTree;
    Drawable imgChosen;
    //性别的默认选项为男,默认称谓为"先生"
    String gender = "先生";
    //定义对话框中某按钮被点击后退出应用的事件方法
    DialogInterface. OnClickListener exitListener = new DialogInterface. OnClickListener(){
        @Override
        public void onClick(DialogInterface dialog,int which){
            MainActivity. this. finish();
        }
```

```java
    };
    @Override
    protected void onCreate(Bundle savedInstanceState){
        super.onCreate(savedInstanceState);
        setContentView(R.layout.activity_main);
        //初始化各个文本框
        edtUserID = (EditText)this.findViewById(R.id.edt_userID);
        edtPss = (EditText)this.findViewById(R.id.edt_pss);
        edtPssConfirm = (EditText)this.findViewById(R.id.edt_pssConfirm);
        //初始化性别单选按钮组
        rdgSex = (RadioGroup)findViewById(R.id.rdg_sex);
        rdgSex.setOnCheckedChangeListener(this);
        //初始化头像,并设置单击监听
        imgBaby = (ImageView)findViewById(R.id.img_baby);
        imgStar = (ImageView)findViewById(R.id.img_star);
        imgTree = (ImageView)findViewById(R.id.img_tree);
        imgBaby.setOnClickListener(this);
        imgStar.setOnClickListener(this);
        imgTree.setOnClickListener(this);
        //将头像默认设置为 imgBaby 所代表图片
        imgChosen = imgBaby.getDrawable();
    }
    @Override
    public void onCheckedChanged(RadioGroup group,int checkedID){
        // 当单选按钮组的选项发生改变时,更改称谓字符串的内容
        if(checkedID == R.id.rd_female){
            gender = "女士";
        }else{
            gender = "先生";
        }
    }
    @Override
    public void onClick(View view){
        // TODO Auto-generated method stub
        switch(view.getId()){
        case R.id.img_baby:
            imgChosen = imgBaby.getDrawable();
            break;
        case R.id.img_star:
            imgChosen = imgStar.getDrawable();
            break;
        case R.id.img_tree:
            imgChosen = imgTree.getDrawable();
            break;
        }
    }
    //单击确定按钮时的事件方法
    public void clickOK(View v){
```

```java
        //判定两次密码是否输入一致
        if(edtPss.getText().toString().equals(edtPssConfirm.getText().toString())){
            //显示个性化的对话框
            showPssCorrectDialog();
        }else{
            //显示密码不一致对话框
            showPssErrorDialog();
        }
    }
    private void showPssCorrectDialog(){
        // 图标是用户选中的头像,内容为
        //"恭喜XX先生/女士(根据性别决定),您获得了完善有礼50元奖券! 祝您消费愉快!"
        StringBuilder content = new StringBuilder()
            .append("恭喜")
            .append(edtUserID.getText().toString())
            .append(gender).append("\n")
            .append("您获得了完善有礼50元奖券! 祝您消费愉快!");
        Dialog pssCorrectDialog = new AlertDialog.Builder(MainActivity.this)
            .setIcon(imgChosen)
            .setTitle("资料成功完善")
            .setMessage(content)
            .setPositiveButton(R.string.ok, exitListener)
            .create();
        pssCorrectDialog.show();
    }
    private void showPssErrorDialog(){
        // TODO Auto-generated method stub
        Dialog pssErrorDialog = new AlertDialog.Builder(MainActivity.this)
            .setIcon(android.R.drawable.ic_dialog_alert)
            .setTitle(R.string.ok)
            .setMessage("两次密码不一致,请重试!")
            .setPositiveButton(R.string.ok, new DialogInterface.OnClickListener(){
                @Override
                public void onClick(DialogInterface dialog, int which){
                    edtPss.setText("");
                    edtPssConfirm.setText("");
                    // 密码框获得焦点
                    edtPss.requestFocus();
                }
            })
            .create();
        pssErrorDialog.show();
    }
    //单击取消时的事件方法
    public void clickCancel(View v){
        showExitDialog();
    }
    private void showExitDialog(){
```

```java
Dialog exitDialog = new AlertDialog.Builder(MainActivity.this)
        .setIcon(android.R.drawable.ic_dialog_info)
        .setTitle(R.string.cancel)
        .setMessage("您确定要取消本次操作吗?")
        .setNegativeButton(R.string.cancel, null)
        .setPositiveButton(R.string.ok, exitListener)
        .create();
exitDialog.show();
    }
}
```

3.4.2 单选按钮组（RadioGroup 与 RadioButton）的定义与使用

1. RadioGroup 与 RadioButton 的定义

在应用程序里，用户可以通过单选按钮完成多选一的操作，RadioGroup 是一个存放单选按钮的容器。此类定义如下：

```
java.lang.Object
  ↳ android.view.View
    ↳ android.view.ViewGroup
      ↳ android.widget.LinearLayout
        ↳ android.widget.RadioGroup
```

在使用这个组件时，一定要在其中配置多个单选按钮才可以使用，此时需要用到 RadioButton 类。此类定义如下：

```
java.lang.Object
  ↳ android.view.View
    ↳ android.widget.TextView
      ↳ android.widget.Button
        ↳ android.widget.CompoundButton
          ↳ android.widget.RadioButton
```

在 res/layout 当中的 XML 文件里注册时可参考如下代码：

```xml
<RadioGroup
    android:id="@+id/rdg_sex"
    android:orientation="horizontal" >
    <RadioButton
        android:id="@+id/rd_male"
        android:layout_width="wrap_content"
        android:layout_height="wrap_content"
        android:drawableRight="@drawable/boy"
        android:checked="true" />
    <RadioButton
        android:id="@+id/rd_female"
        android:layout_width="wrap_content"
        android:layout_height="wrap_content"
```

```
                    android:drawableRight = "@drawable/girl"/>
    </RadioGroup>
```

从 RadioGroup 的定义中可以看出,它继承于 LinearLayout,所以在上述代码中,可以使用 android:orientation 属性设置其内部子组件的排列方向。

使用 RadioButton 的 android:drawableRight 属性可以在单选钮的右侧放置图片资源。相应地,如果希望将图片放在单选钮的左侧,则可使用 android:drawableLeft,其余方向,依此类推。为了图片资源可以正确访问,需要提前将准备好的图片存放在 res/drawable-x 文件夹中。

与复选框相似,RadioButton 的 android:checked 属性可以设置其初始状态是否被选中,值为 true 或 false。

2. RadioGroup 与 RadioButton 的常用方法

RadioButton 与 CheckBox 都是继承于 CompoundButton,二者常用的方法完全相同,具体内容可参考本章第 3.3.3 节。

RadioGroup 的常用方法如表 3-13 所示。

表 3-13　RadioGroup 的常用方法说明

序号	方法名称	说明
1	public void setOnCheckedChangeListener(RadioGroup.OnCheckedChangeListener listener)	注册 RadioGroup 中单选钮选中状态发生改变的监听
2	public int getCheckedRadioButtonId()	返回 RadioGroup 中被选中的单选钮 id
3	public void check(int id)	选中指定 id 的单选钮
4	public abstract void onCheckedChanged(RadioGroup group, int checkedId)	在此方法中实现当 RadioGroup 中单选钮选中状态发生改变时的具体处理工作

3.4.3　图像框(ImageView)的定义与使用

1. ImageView 的定义

图像框 ImageView 的主要功能是为图片展示提供一个容器。此类定义如下:

```
    java.lang.Object
    ↳ android.view.View
      ↳ android.widget.ImageView
```

在使用这个组件时,首先在 res/layout 当中的 xml 文件里进行注册,注册时可参考如下代码:

```
    <ImageView
        android:id = "@+id/img_baby"
        android:layout_width = "wrap_content"
        android:layout_height = "wrap_content"
        android:src = "@drawable/baby"/>
```

在 XML 文件中,可以使用 android:src 属性设置 ImageView 所显示的图片资源,为了使用方便,需要将准备好的图片存放在 res/drawable-x 文件夹中。

2. ImageView 的常用方法

从类的定义中可以看出，ImageView 类是 View 类的子类，所以其常用方法中有一些继承于 View，如表 3-14 所示。

表 3-14　ImageView 的常用方法说明

序号	方法名称	说明
1	public void setOnClickListener(View.OnClickListener l)	继承于 View 类的方法，为单击事件注册监听
2	public abstract void onClick(View v)	在此方法中实现单击时的具体处理工作
3	public Drawable getDrawable()	以 Drawable 为类型取得 ImageView 的图片
4	public void setImageResource(int resId)	为 ImageView 设置图片资源，参数为图片资源 ID
5	public void setImageBitmap(Bitmap bm)	为 ImageView 填充图片，参数为 Bitmap 类型
6	public void setImageDrawable(Drawable drawable)	为 ImageView 填充图片，参数为 Drawable 类型

在实例 4 中，为了判断用户到底选择了哪个图像作为其头像，首先让 MainActivity 类实现了 OnClickListener 接口，代码如下：

```
public class MainActivity extends Activity implements OnClickListener
```

其次在 onCreate() 方法中为 3 个图像框注册了单击监听，代码如下：

```
imgBaby.setOnClickListener(this);
imgStar.setOnClickListener(this);
imgTree.setOnClickListener(this);
```

然后，对 3 个图像框如何响应单击事件做了统一处理，代码如下：

```
@Override
public void onClick(View view){
    //首先判断用户单击了哪个图像框，然后将图像框的图片赋值给 imgChosen
    switch(view.getId()){
        case R.id.img_baby:
            imgChosen = imgBaby.getDrawable();
            break;
        case R.id.img_star:
            imgChosen = imgStar.getDrawable();
            break;
        case R.id.img_tree:
            imgChosen = imgTree.getDrawable();
            break;
    }
}
```

这样设计可以使程序代码整洁，层次结构清晰，便于其他开发人员阅读和维护。

3.4.4　警告对话框（AlertDialog 与 AlertDialog.Builder）使用简介

1. AlertDialog 与 AlertDialog.Builder 的定义

警告对话框 AlertDialog 是各种应用中出现的最频繁、最简单的一种对话框，主要目的是

为用户显示一条警告信息。此类定义如下:

> java. lang. Object
> ↳ android. app. Dialog
> ↳ android. app. AlertDialog

在使用时,首先需要依靠其内部类 AlertDialog. Builder 来实例化 AlertDialog 类。AlertDialog. Builder 类的定义如下:

> java. lang. Object
> ↳ android. app. AlertDialog. Builder

实例化时需要用到 AlertDialog. Builder 类的构造方法 AlertDialog. Builder(Context context),其中参数为当前环境,参考代码如下:

> DialogmyDialog = new AlertDialog. Builder(MainActivity. this);

2. AlertDialog. Builder 的常用方法

AlertDialog. Builder 类常用的方法如表 3-15 所示。

表 3-15 <ImageButton> 的常用方法说明

序号	方法名称	说明
1	public AlertDialog create( )	创建 AlertDialog 的实例化对象
2	public AlertDialog. Builder setIcon(Drawable icon)	设置对话框的显示图标,参数可以是 Drawable 类型,也可以是图标资源 id,但必须与 setTitle 一起使用
3	public AlertDialog. Builder setIcon(int iconId)	
4	public AlertDialog. Builder setItems(intitemsId, DialogInterface. OnClickListener listener)	在对话框中设置列表项,列表项内容可以来自于数组资源文件,也可以是字符串数组,分别通过资源 id 和数组变量 items 来传递,同时设置监听 listener
5	public AlertDialog. Builder setItems(CharSequence[ ] items, DialogInterface. OnClickListener listener)	
6	public AlertDialog. Builder setMessage(CharSequence message)	设置对话框的消息,参数可以是字符串资源 id,也可以是字符串变量
7	public AlertDialog. Builder setMessage(int messageId)	
8	public AlertDialog. Builder setNegativeButton(inttextId, DialogInterface. OnClickListener listener)	为对话框设置取消按钮,第 1 个参数是按钮上的显示文字,可以是字符串资源 id,也可以是字符串变量 text;第 2 个参数是为按钮设置的监听操作
9	public AlertDialog. Builder setNegativeButton(CharSequence text, DialogInterface. OnClickListener listener)	
10	public AlertDialog. Builder setNeutralButton(inttextId, DialogInterface. OnClickListener listener)	为对话框设置普通按钮,第 1 个参数是按钮上的显示文字,可以是字符串资源 id,也可以是字符串变量 text;第 2 个参数是为按钮设置的监听操作
11	public AlertDialog. Builder setNeutralButton(CharSequence text, DialogInterface. OnClickListener listener)	
12	public AlertDialog. Builder setPositiveButton(inttextId, DialogInterface. OnClickListener listener)	为对话框设置确定按钮,第 1 个参数是按钮上的显示文字,可以是字符串资源 id,也可以是字符串变量 text;第 2 个参数是为按钮设置的监听操作
13	public AlertDialog. Builder setPositiveButton(CharSequence text, DialogInterface. OnClickListener listener)	

(续)

序号	方法名称	说明
14	public AlertDialog.Builder setTitle(CharSequence title)	为对话框设置标题，参数可以是字符串变量 title，也可以是字符串资源的 id
15	public AlertDialog.Builder setTitle(int titleId)	
16	public AlertDialog.Builder setOnItemSelectedListener(AdapterView.OnItemSelectedListener listener)	为对话框设置当数据项被选中时的监听
17	public AlertDialog.Builder setView(View view)	为对话框设置个性化的 view 组件
18	public void show()	继承于 Dialog 类的方法，用于启动对话框，并在屏幕上显示

3.5 动手实践 3：为友秀宝

3.5.1 功能要求

1. 应用程序首界面

应用程序的首界面如图 3-6 所示。

2. 各个组件的设置要求

用户可以在"宝贝称呼"对应的编辑框中只能输入文本信息，在"宝贝年龄"对应的编辑框中只能输入数字，效果如图 3-7 所示。

用户可以从下拉列表框中选择"祖籍"信息，如图 3-8 所示。

图 3-6 应用程序首界面　　　图 3-7 输入称呼和生日信息　　　图 3-8 祖籍列表框

用户可以通过单击单选钮选择"宝贝性别"，默认为"男"；通过单击图像框确定"宝贝长相"，默认长相为"大象"图标。

3. 为好友发送"宝贝"信息

当用户单击"秀"按钮时，如果宝贝的所有信息已经填写完整，则弹出如图 3-9 所示的"秀"对话框。单击对话框中的"秀"按钮后发送信息，同时应用程序结束运行。

接收方看到的信息页面如图 3-10 所示。

图 3-9　为友人秀宝对话框　　　　　　　图 3-10　接收方看到的短信

当用户单击"秀"按钮时，如果宝贝的信息尚未填写完整，则给出如图 3-11 所示的 Toast 提示，并让不完整的组件获得焦点，如图 3-11 所示。

当用户单击"走"按钮时，给出如图 3-12 所示的对话框，在对话框中单击"确定"退出应用，单击"取消"返回应用主页面。

图 3-11　信息不完整时的提示　　　　　　图 3-12　单击"走"按钮效果

3.5.2　操作提示

1. 定义布局

应用程序首页面在 res/layout/activity_main.xml 中保存，可以参考实例 4 完成。部分代码参考如下：

```xml
    ...
            <!-- 只能输入数字的年龄编辑框 -->
            <EditText
                android:id = "@+id/edt_age"
                android:layout_width = "wrap_content"
                android:layout_height = "wrap_content"
                android:layout_marginLeft = "28dp"
                android:inputType = "number"/>
```

```xml
...
    <TableRow>
        <TextView
            android:id="@+id/tv_image"
            android:layout_width="wrap_content"
            android:layout_height="wrap_content"
            android:text="@string/image"
            android:layout_gravity="right|center_vertical"/>
        <!--嵌套线性布局,将4个图像框水平方向依次排列-->
        <LinearLayout
            android:orientation="horizontal"
            android:layout_width="wrap_content"
            android:layout_height="wrap_content"
            android:layout_marginLeft="28dp"
            android:gravity="center_horizontal">
            <!--为4个图像框设置相同的onClick事件方法,便于在Java中统一处理-->
            <ImageView
                android:id="@+id/img_dragon"
                android:layout_width="wrap_content"
                android:layout_height="wrap_content"
                android:src="@drawable/dragon"
                android:layout_margin="1dp"
                android:onClick="clickIMG"/>
            <ImageView
                android:id="@+id/img_pica"
                android:layout_width="wrap_content"
                android:layout_height="wrap_content"
                android:src="@drawable/pica"
                android:layout_margin="1dp"
                android:onClick="clickIMG"/>
            <ImageView
                android:id="@+id/img_cat"
                android:layout_width="wrap_content"
                android:layout_height="wrap_content"
                android:src="@drawable/cat"
                android:layout_margin="1dp"
                android:onClick="clickIMG"/>
            <ImageView
                android:id="@+id/img_elephant"
                android:layout_width="wrap_content"
                android:layout_height="wrap_content"
                android:src="@drawable/elephant"
                android:layout_margin="1dp"
                android:onClick="clickIMG"/>
        </LinearLayout>
    </TableRow>
...
    <!--使用android:drawableLeft属性,在按钮左侧添加图片-->
```

```xml
<Button
    android:id = "@+id/btn_ok"
    android:layout_width = "wrap_content"
    android:layout_height = "wrap_content"
    android:text = "@string/ok"
    android:drawableLeft = "@drawable/show"
    android:onClick = "clickOK"/>
```

2. 为图像框添加单击事件

在 res/layout/activity_main.xml 布局中，4个图像框的 onClick 属性值完全相同，便于在 MainActivity.java 中统一处理。参考代码如下：

```java
public void clickIMG(View v){
    switch(v.getId()){
    case R.id.img_cat:
        imgChosen = imgCat.getDrawable();
        break;
    case R.id.img_dragon:
        imgChosen = imgDragon.getDrawable();
        break;
    case R.id.img_elephant:
        imgChosen = imgElephant.getDrawable();
        break;
    case R.id.img_pica:
        imgChosen = imgPica.getDrawable();
        break;
    }
}
```

上述代码中 imgChosen 是 MainActivity 类中 Drawable 类型的成员变量。

3. 实现发送信息

图3-9中在对话框中使用了定制的布局 sendView。参考代码如下：

```java
//使用自定义的 sendView 作为对话框的视图
Dialog sendDialog = new AlertDialog.Builder(MainActivity.this)
    .setIcon(imgChosen)
    .setTitle(R.string.ok)
    .setView(sendView)
    .setPositiveButton(R.string.ok, clickSendListener)
    .create();
```

sendView 是 MainActivity.java 类中的成员变量，用 res/layout/sendlayout.xml 布局实现初始化。初始化任务在 MainActivity 的 onCreate() 方法中完成。参考代码如下：

```java
View sendView;
...
protected void onCreate(Bundle savedInstanceState){
    ...
    //使用定制的 sendlayout 作为对话框的视图 sendView
```

```
sendView = LayoutInflater.from(MainActivity.this).inflate(R.layout.sendlayout,null);
//取出 sendView 中的电话号码
edtPhone = (EditText)sendView.findViewById(R.id.edt_phone);
...
}
```

res/layout/sendlayout.xml 中包含了水平方向排列的 1 个 TextView 和 1 个 EditText。参考代码如下：

```xml
<?xml version = "1.0" encoding = "utf-8"?>
<LinearLayout xmlns:android = "http://schemas.android.com/apk/res/android"
    android:layout_width = "match_parent"
    android:layout_height = "match_parent"
    android:orientation = "horizontal" >
<TextView
    android:id = "@+id/textView1"
    android:layout_width = "wrap_content"
    android:layout_height = "wrap_content"
    android:text = "为他秀" />
<EditText
    android:id = "@+id/edt_phone"
    android:layout_width = "match_parent"
    android:layout_height = "wrap_content"
    android:inputType = "phone" >
<requestFocus />
</EditText>
</LinearLayout>
```

使用"sendTextMessage()"方法发送短信时，需要注意保证电话号码和短信内容不为空，其次保证短信内容不超过 70 个字符的限制，如果超过要对短信内容进行分割处理。

第4章 Android 中的高级视图组件

第3章介绍了 Android 提供的基本界面组件，本章向大家介绍其他常用的高级组件。使用这些组件可以大大降低开发者的开发难度，为快速程序开发提供方便。

4.1 实例1：随心换肤

4.1.1 功能要求与操作步骤

1. 功能要求

完成如图 4-1 所示的应用程序。要求能够：

图 4-1 "随心换肤"运行效果图

① 在图 4-1a、图 4-1b 所示的界面中，"上一幅"按钮和"下一幅"按钮分别用来显示预先准备好的图片数组中的图片。当单击这两个按钮时，图片位置的提示文本也会随之发生改变。

② 当显示到数组中最后一张图片并且单击"下一幅"按钮时，从头开始循环显示；当

显示到数组的第一张图片并且单击"上一幅"按钮时,从最后一幅开始循环显示。

③ 当单击4-1中的图片时,可以改变手机桌面背景并给出提示,如图4-1c、图4-1d所示。

2. 操作步骤

(1) 创建项目

参考如下输入信息创建 Android 项目:

> Application Name:04_TestViewSwitcher
> Project Name:04_TestViewSwitcher
> Package Name:com.book.testviewswitcher
> Activity Name:MainActivity
> Layout Name:activity_main

(2) 定义布局

从图4-1中可看出,本应用可以使用嵌套线性布局来实现,各个组件的属性可参考表4-1设置。

表4-1 实例1中各组件的属性列表

序号	组件类型	属性名称	属性值	说明
1	LinearLayout	android:layout_width	match_parent	宽度匹配父容器
		android:layout_height	match_parent	高度匹配父容器
		android:orientation	vertical	垂直方向
2	ImageSwitcher	android:id	@+id/img_show	增加 id 为"img_show"的组件
		android:layout_width	match_parent	宽度匹配父容器
		android:layout_height	0dp	高度为0dp,与权重配合使用
		android:layout_weight	8	占整个布局的80%
		android:onClick	clickImage	单击事件为"clickImage"
3	TextSwitcher	android:id	@+id/txt_show	增加 id 为"txt_show"的组件
		android:layout_width	match_parent	宽度匹配父容器
		android:layout_height	0dp	高度为0dp,与权重配合使用
		android:layout_weight	1	占整个布局的10%
		android:background	@android:color/black	背景为黑色
4	LinearLayout	android:layout_width	match_parent	宽度匹配父容器
		android:layout_height	0dp	高度为0dp,与权重配合使用
		android:orientation	horizontal	水平方向
		android:layout_weight	1	占整个布局的10%
		android:gravity	center	子组件居中对齐
5	Button(两个)	android:layout_width	wrap_content	宽度适合内容
		android:layout_height	wrap_content	高度适合内容
		android:text	上一幅/下一幅	显示文本"上一幅"和"下一幅"
		android:onClick	clickPrev/clickNext	单击事件为"clickPrev"和"clickNext"

109

参考代码如下:

```xml
<LinearLayout xmlns:android = "http://schemas.android.com/apk/res/android"
    xmlns:tools = "http://schemas.android.com/tools"
    android:layout_width = "match_parent"
    android:layout_height = "match_parent"
    android:orientation = "vertical" >
    <ImageSwitcher
        android:id = "@+id/img_show"
        android:layout_width = "match_parent"
        android:layout_height = "0dp"
        android:layout_weight = "8"
        android:onClick = "clickImage" >
    </ImageSwitcher>
    <TextSwitcher
        android:id = "@+id/txt_show"
        android:layout_width = "match_parent"
        android:layout_height = "0dp"
        android:layout_weight = "1"
        android:background = "@android:color/black"
        />
    <LinearLayout
        android:layout_width = "match_parent"
        android:layout_height = "0dp"
        android:orientation = "horizontal"
        android:layout_weight = "1"
        android:gravity = "center" >
        <Button
            android:id = "@+id/button1"
            android:layout_width = "wrap_content"
            android:layout_height = "wrap_content"
            android:text = "上一幅"
            android:onClick = "clickPrev" />
        <Button
            android:id = "@+id/button2"
            android:layout_width = "wrap_content"
            android:layout_height = "wrap_content"
            android:text = "下一幅"
            android:onClick = "clickNext" />
    </LinearLayout>
</LinearLayout>
```

(3) 准备图片资源

为了丰富图片浏览效果,本例中将使用到8幅图片,名称为"pic01.jpg""pic02.jpg""pic03.jpg""pic04.jpg""pic05.jpg""pic06.jpg""pic07.jpg""pic08.jpg",存放在res/drawable-x目录中,在Java文件中将其id依次放在int类型的数组中即可。

(4) 实现MainActivity中的功能

在本应用中,单击"上一幅"按钮和"下一幅"按钮时都需要实现按钮的单击事件

clickPrev 和 clickNext 来达到显示上一幅图片或下一幅图片，以及更改提示文本的目的。当单击图片切换器组件时，需要触发 clickImage 事件，以便设置手机桌面背景。

参考代码如下：

```java
public class MainActivity extends Activity {
    //定义图片切换器对象 imgShow
    ImageSwitcher imgShow;
    //定义文本切换器对象 txtShow
    TextSwitcher txtShow;
    //用于修改桌面背景的图片 bitmap
    Bitmap bitmap = null;
    //定义图片数组
    int [ ]imgRes = new int[ ]{R. drawable. pic01,
            R. drawable. pic02, R. drawable. pic03,
            R. drawable. pic04, R. drawable. pic05,
            R. drawable. pic06, R. drawable. pic07,
            R. drawable. pic08};
    //定义文本数组
    String [ ] txtRes = new String[ ]{
            "第1幅","第2幅","第3幅","第4幅",
            "第5幅","第6幅","第7幅","第8幅"
    };
    //定义数组下标
    int index = 0;
    @Override
    protected void onCreate(BundlesavedInstanceState) {
        super. onCreate(savedInstanceState);
        setContentView(R. layout. activity_main);
        imgShow = (ImageSwitcher)this. findViewById(R. id. img_show);
        txtShow = (TextSwitcher)this. findViewById(R. id. txt_show);
        //定义视图操作工厂对象 imgFactory
        ImageViewFactory imgFactory = new ImageViewFactory();
        imgShow. setFactory(imgFactory);
        //定义视图操作工厂对象 txtFactory
        TextViewFactory txtFactory = new TextViewFactory();
        txtShow. setFactory(txtFactory);
        //为图片切换器设置切换的动画效果
        imgShow. setInAnimation(AnimationUtils. loadAnimation(this,
                                android. R. anim. fade_in));
        imgShow. setOutAnimation(AnimationUtils. loadAnimation(this,
                                android. R. anim. fade_out));
        //为图片切换器设置图片
        imgShow. setImageResource(imgRes[index]);
        //为文本切换器设置切换的动画效果
        txtShow. setInAnimation(AnimationUtils. loadAnimation(this,
                                android. R. anim. fade_in));
        txtShow. setOutAnimation(AnimationUtils. loadAnimation(this,
                                android. R. anim. fade_out));
```

```java
        //为文本切换器设置文本
        txtShow.setText("当前图片位置:" + txtRes[index]);
        //将当前显示图片解码为bitmap对象
        bitmap = BitmapFactory.decodeResource(getResources(), imgRes[index]);
}
//图片切换器的单击事件
public void clickImage(View v){
        //设置手机桌面背景,别忘了在AndroidManifest.xml中添加权限
        // <uses-permission android:name="android.permission.SET_WALLPAPER"/>
        try{
        //MainActivity.this.setWallpaper(bitmap);
        //替代方法如下:
        WallpaperManager wallPaper = WallpaperManager.getInstance(MainActivity.this);
        wallPaper.setBitmap(bitmap);
        } catch(IOException e){
            // TODO Auto-generated catch block
            e.printStackTrace();
        }
        Toast.makeText(MainActivity.this,"手机桌面已经更换,快去看看吧!",
                Toast.LENGTH_LONG).show();
}
//上一幅的单击事件
public void clickPrev(View v){
        //如果已经浏览到图片数组的第一幅,则将数组下标定位到最后一幅,以便循环显示
        if(index <= 0){
            index = imgRes.length;
        }
        imgShow.setImageResource(imgRes[--index]);
        txtShow.setText("当前图片位置:" + txtRes[index]);
        bitmap = BitmapFactory.decodeResource(getResources(), imgRes[index]);
}
//下一幅的单击事件
public void clickNext(View v){
        //如果已经浏览到图片数组的最后一幅,则将数组下标定位到第一幅,以便循环显示
        if(index >= imgRes.length - 1){
            index = -1;
        }
        imgShow.setImageResource(imgRes[++index]);
        txtShow.setText("当前图片位置:" + txtRes[index]);
        bitmap = BitmapFactory.decodeResource(getResources(), imgRes[index]);
}
//定义视图操作工厂类
public class ImageViewFactory implements ViewFactory{
        @Override
        public View makeView(){
            // TODO Auto-generated method stub
            ImageView img = new ImageView(MainActivity.this);
            img.setScaleType(ImageView.ScaleType.FIT_XY);
```

```
                    img.setLayoutParams(new ImageSwitcher.LayoutParams(
                        LayoutParams.MATCH_PARENT,
                        LayoutParams.MATCH_PARENT));
                    return img;
                }
            }
            public class TextViewFactory implements ViewFactory {
                @Override
                public View makeView() {
                    // TODO Auto-generated method stub
                    TextView text = new TextView(MainActivity.this);
                    //设置文本颜色
                    text.setTextColor(Color.WHITE);
                    text.setLayoutParams(new TextSwitcher.LayoutParams(
                        LayoutParams.MATCH_PARENT,
                        LayoutParams.MATCH_PARENT));
                    //设置文本大小
                    text.setTextSize(30);
                    //设置文本对齐方式
                    text.setGravity(Gravity.CENTER);
                    return text;
                }
            }
        }
```

（5）在 AndroidManifest.xml 文件中添加权限

为了能够使应用程序设置手机桌面背景，需要在 AndroidManifest.xml 文件中添加相关权限，参考代码如下：

```
<uses-permission android:name="android.permission.SET_WALLPAPER"/>
```

4.1.2　图片切换器（ImageSwitcher）的定义与使用

1. ImageSwitcher 的定义

图片切换器 ImageSwitcher 的主要功能是完成图片的切换显示，例如用户在进行图片浏览时，可以通过单击按钮逐张切换显示的图片，而且使用 ImageSwitcher 组件切换图片时，还可以为其增加一些动画效果。此类定义如下：

```
java.lang.Object
  ↳ android.view.View
    ↳ android.view.ViewGroup
      ↳ android.widget.FrameLayout
        ↳ android.widget.ViewAnimator
          ↳ android.widget.ViewSwitcher
            ↳ android.widget.ImageSwitcher
```

在使用这个组件时，首先在 res/layout 中的 XML 文件里进行注册，注册时可参考如下代码：

```
< ImageSwitcher
    android:id = "@+id/img_show"
    android:layout_width = "match_parent"
    android:layout_height = "match_parent"
    android:onClick = "clickImage" >
</ImageSwitcher >
```

2. ImageSwitcher 的常用方法

从类的定义中可以看出,ImageSwitcher 类继承于 View、ViewGroup、FrameLayout、ViewAnimator、ViewSwitcher 等类,所以其常用方法中有一些继承于这些类,如表 4-2 所示。

表 4-2 ImageSwitcher 的常用方法说明

序号	方法名称	说明
1	public void setFactory(ViewSwitcher.ViewFactory factory)	继承于 ViewSwitcher 类的方法,设置 ViewFactory 对象,用于完成两个图片切换时 ViewSwitcher 的转换操作
2	public void setImageResource(int resid)	设置显示的图片资源 ID
3	public void setInAnimation(Animation inAnimation)	继承于 ViewAnimator 的方法,用于设置图片进入到 ImageSwitcher 时的动画效果
4	public void setOutAnimation(Animation inAnimation)	继承于 ViewAnimator 的方法,用于设置图片消失于 ImageSwitcher 时的动画效果

3. ImageSwitcher 动画效果的实现方法

从类的定义中可以看出,ImageSwitcher 继承于 ViewAnimator,调用该类的 setInAnimation(Animation in) 和 setOutAnimation(Animation out) 方法即可实现动画效果。

这两个方法中的参数为 Animation 类的对象,用于指定动画类型。此类定义如下:

```
java.lang.Object
  ↳ android.view.animation.Animation
```

但设置动画效果时,需要首先取得 Animation 类的对象,再借助 AnimationUtils 类完成。此类的定义如下:

```
java.lang.Object
  ↳ android.view.animation.AnimationUtils
```

使用该类的 loadAnimation(Context c,int id) 方法可以完成动画设置,只是需要指定动画的资源类型,这些类型可以直接从 android.R 中定义的常量找出,本例中使用到了两个资源常量,如表 4-3 所示。

表 4-3 android.R 定义的动画资源常量

序号	常量名称	说明
1	public static final int fade_in	"淡入"动画
2	public static final int fade_out	"淡出"动画

4. ImageSwitcher 实现图片切换功能

如果需要实现图片切换的效果,还必须要在定义的 Activity 类中实现 ViewSwitcher.ViewFactory 接口,以指定切换视图的操作工厂,此接口定义如下:

```
public static interface ViewSwitcher.ViewFactory {
    //创建一个新的 View 对象,将其加入到 ViewSwitcher 中,并返回新的 View 对象
    public abstract View makeView( );
}
```

4.1.3 文本切换器（TextSwitcher）的定义与使用

1. TextSwitcher 的定义

与图片切换器相似，文本切换器 TextSwitcher 的主要功能是完成文本的切换显示。例如在本例中，用户在进行图片浏览时，可以根据显示图片的不同而不断地更新其提示文本。使用 TextSwitcher 组件切换文本时，也可以为其增加一些动画效果。此类定义如下：

```
java.lang.Object
  ↳ android.view.View
    ↳ android.view.ViewGroup
      ↳ android.widget.FrameLayout
        ↳ android.widget.ViewAnimator
          ↳ android.widget.ViewSwitcher
            ↳ android.widget.TextSwitcher
```

在使用这个组件时，首先在 res/layout 中的 XML 文件里进行注册，注册时可参考如下代码：

```
<TextSwitcher
    android:id = "@ + id/txt_show"
    android:layout_width = "match_parent"
    android:layout_height = "0dp"
    android:layout_weight = "1"
    android:background = "@android:color/black"
/>
```

2. TextSwitcher 的使用方法

与 ImageSwitcher 相似，文本切换器的常用方法如表 4-4 所示。

表 4-4 < TextSwitcher > 的常用方法说明

序号	方法名称	说明
1	public void setFactory(ViewSwitcher.ViewFactory factory)	继承于 ViewSwitcher 类的方法，设置 ViewFactory 对象，用于完成两个图片切换时 ViewSwitcher 的转换操作
2	public void setText(CharSequence text)	设置显示的文本
3	public void setInAnimation(Animation inAnimation)	继承于 ViewAnimator 的方法，用于设置图片进入到 ImageSwitcher 时的动画效果
4	public void setOutAnimation(Animation inAnimation)	继承于 ViewAnimator 的方法，用于设置图片消失于 ImageSwitcher 时的动画效果

与图片切换器相同，调用 ViewAnimator 类的 setInAnimation(Animation in) 和 setOutAnimation(Animation out) 方法即可实现动画效果。如果需要实现文本切换的效果，也必须要在定义的 Activity 类中实现 ViewSwitcher.ViewFactory 接口，以指定切换视图的操作工厂，具体

的实现方法参考上一节。

4.1.4 设置手机桌面背景简介

1. 设置手机桌面背景的方法

设置手机桌面背景使用系统的 setWallpaper(Bitmap bitmap) 方法,其中参数为 Bitmap 对象。当调用该方法时,将手机桌面背景设置为已选定的图片。

参考代码为:

```
MainActivity.this.setWallpaper(bitmap);
```

也可以通过获取 WallpaperManager 实例对象,再调用其 setBitmap(Bitmap bitmap) 方法来设置。

参考代码为:

```
WallpaperManager wallPaper = WallpaperManager.getInstance(MainActivity.this);
wallPaper.setBitmap(bitmap);
```

上述两段参考代码中 MainActivity.this 均为当前的 Activity 类。

2. 设置权限

初学者很容易忽视权限的问题,再次强调,如果需要应用程序能设置手机桌面背景,那么一定要在 AndroidManifest.xml 文件中添加以下权限:

```
<uses-permission android:name="android.permission.SET_WALLPAPER"/>
```

4.2 实例2: 居家助手

4.2.1 功能要求与操作步骤

1. 功能要求

完成如图 4-2 所示的应用程序。要求能够:

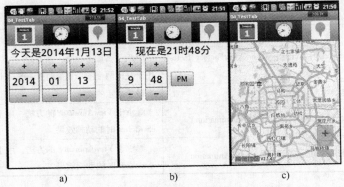

图 4-2 "居家助手"运行效果图

① 在图 4-2 所示的各个界面中,共有 3 个选项卡,分别可以用来显示日期、时间和地图。

② 单击各个选项卡标签可以分别在 3 个选项卡之间进行切换。

③ 在图 4-2a 中，当前系统日期用 30sp 的文本显示；在图 4-2b 中，应用程序运行时的系统时间用 30sp 的文本显示。

注：为了能够判断地图页是否可以正常显示，最好使用真机测试。

2. 操作步骤

（1）创建项目

参考如下输入信息创建 Android 项目：

> Application Name：04_TestTab
> Project Name：04_TestTab
> Package Name：com. book. testtab
> Activity Name：MainActivity
> Layout Name：activity_main

（2）获取高德地图 SDK，申请 API Key

从网站（http://developer.amap.com/）下载高德地图 Android SDK。下载界面如图 4-3 所示。

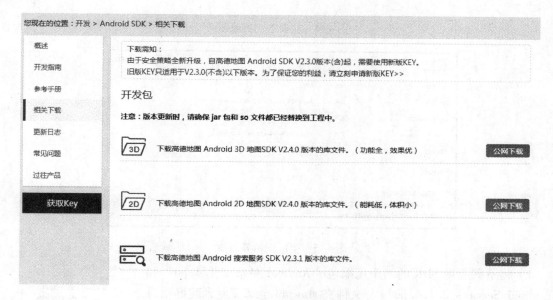

图 4-3　下载地图 SDK 界面图

然后，按照网站相关流程获取 API 密钥。注册密钥的界面如图 4-4 所示。

在输入应用名称（注意：在为本例申请密钥时，需将应用名称填写为"04_TestTab"。在为其他应用程序申请密钥时，需要将应用名称修改）和正确的验证码之后，单击图 4-4 中的"获取 KEY"按钮即可在按钮下方看到"注册成功，请到我的密钥查看"的提示信息。

单击"我的密钥"超链接即可显示申请的密钥，如图 4-5 所示。

如果需要再次生成密钥，可以单击图 4-5 右上角的"获取 KEY"超链接，再次打开图 4-4 的注册密钥界面，完成相关操作。

（3）配置项目

下载好 SDK 并生成密钥后，可以按照以下步骤配置项目。

图 4-4 注册密钥界面图

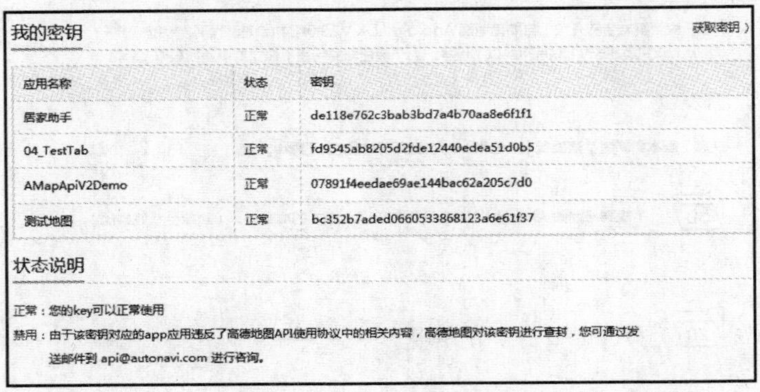

图 4-5 生成密钥界面图

1)将下载所得 SDK 的两个压缩包"Android_Map_2.1.4.jar"和"Android_Services_2.1.4.jar"与文件夹 armeabi 全部复制到建好项目的 libs 文件夹中。

此时项目的结构如图 4-6 所示。

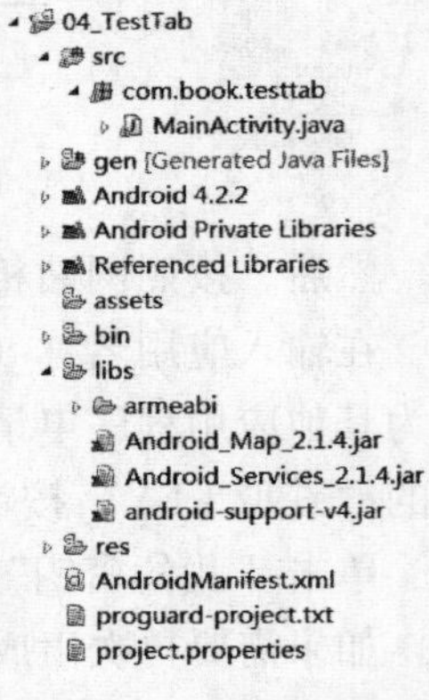

2)选中项目,在右键菜单中,单击"Build Path"命令,然后在弹出的子菜单中选中"Configure Build Path…",打开如图 4-7a 的对话框,查看两个 jar 包是否均已在 Libraries 中包含。

3)在确定两个 jar 包都已经包含在 Libraries 中之后,单击图 4-7a 中的"Order and Export"选项卡,将所有的 jar 包都选中,如图 4-7b 所示,最后单击"OK"按钮关闭对话框。

4)在项目的"AndroidManifest.xml"中添加用户 Key。

参考代码如下:

图 4-6 项目结构图

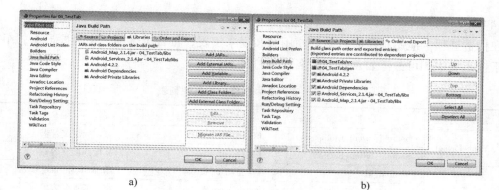

图4-7 Configure Build Path 对话框

```xml
<!--应用信息-->
<application
    android:allowBackup = "true"
    android:icon = "@drawable/ic_launcher"
    android:label = "@string/app_name"
    android:theme = "@style/AppTheme" >
    <!--添加密钥-->
    <meta-data
        android:name = "com.amap.api.v2.apikey"
        android:value = "fd9545ab8205d2fde12440edea51d0b5"
    />
    <!--Activity的配置-->
    <activity
        android:name = "com.book.testtab.MainActivity"
        android:label = "@string/app_name" >
        <intent-filter>
            <action android:name = "android.intent.action.MAIN" />
            <category android:name = "android.intent.category.LAUNCHER" />
        </intent-filter>
    </activity>
</application>
```

5）在项目的"AndroidManifest.xml"中添加权限。

参考代码如下：

```xml
<!--添加权限-->
<uses-permission android:name = "android.permission.INTERNET" />
<uses-permission android:name = "android.permission.WRITE_EXTERNAL_STORAGE" />
<uses-permission android:name = "android.permission.ACCESS_COARSE_LOCATION" />
<uses-permission android:name = "android.permission.ACCESS_NETWORK_STATE" />
<uses-permission android:name = "android.permission.ACCESS_FINE_LOCATION" />
<uses-permission android:name = "android.permission.READ_PHONE_STATE" />
<uses-permission android:name = "android.permission.CHANGE_WIFI_STATE" />
<uses-permission android:name = "android.permission.ACCESS_WIFI_STATE" />
<uses-permission android:name = "android.permission.CHANGE_CONFIGURATION" />
```

（4）定义布局

在构造选项卡应用程序时，最好使用帧布局来实现，将各个选项卡中需要显示的组件依次置于帧布局中。可参考表4-5设置各个组件的属性。

表4-5 实例2中各组件的属性列表

序号	组件类型	属性名称	属性值	说明
1	FrameLayout	android:layout_width	fill_parent	宽度匹配父容器
		android:layout_height	fill_parent	高度匹配父容器
2	LinearLayout	android:id	@+id/date_layout	增加id为"date_layout"的组件
		android:layout_width	match_parent	宽度匹配父容器
		android:layout_height	match_parent	高度匹配父容器
		android:orientation	vertical	垂直方向
3	TextView	android:id	@+id/tv_date	增加id为"tv_date"的组件
		android:layout_width	wrap_content	宽度适合内容即可
		android:layout_height	wrap_content	高度适合内容即可
		android:textSize	30sp	字号为30sp
		android:textColor	@android:color/black	字的颜色为黑色
		android:layout_gravity	center_horizontal	文字居中对齐
4	DatePicker	android:id	@+id/show_date	增加id为"show_date"的组件
		android:layout_width	wrap_content	宽度适合内容即可
		android:layout_height	wrap_content	高度适合内容即可
5	LinearLayout	android:id	@+id/time_layout	增加id为"time_layout"的组件
		android:layout_width	match_parent	宽度匹配父容器
		android:layout_height	match_parent	高度匹配父容器
		android:orientation	vertical	垂直方向
6	TextView	android:id	@+id/tv_time	增加id为"tv_time"的组件
		android:layout_width	wrap_content	宽度适合内容即可
		android:layout_height	wrap_content	高度适合内容即可
		android:textSize	30sp	字号为30sp
		android:textColor	@android:color/black	字的颜色为黑色
		android:layout_gravity	center_horizontal	文字居中对齐
7	TimePicker	android:id	@+id/show_time	增加id为"show_time"的组件
		android:layout_width	wrap_content	宽度适合内容即可
		android:layout_height	wrap_content	高度适合内容即可
8	com.amap.api.maps.MapView	android:id	@+id/map	增加id为"map"的组件
		android:layout_width	match_parent	宽度匹配父容器
		android:layout_height	match_parent	高度匹配父容器

参考代码如下：

```
<?xml version = "1.0" encoding = "utf-8"?>
<FrameLayout xmlns:android = "http://schemas.android.com/apk/res/android"
    android:layout_width = "fill_parent"
    android:layout_height = "fill_parent" >
```

```xml
<LinearLayout
    android:id="@+id/date_layout"
    android:layout_width="match_parent"
    android:layout_height="match_parent"
    android:orientation="vertical"
    >
<TextView
    android:id="@+id/tv_date"
    android:layout_width="wrap_content"
    android:layout_height="wrap_content"
    android:textSize="30sp"
    android:textColor="@android:color/black"
    android:layout_gravity="center_horizontal"
    />
<DatePicker
    android:id="@+id/show_date"
    android:layout_width="wrap_content"
    android:layout_height="wrap_content"
    />
</LinearLayout>
<LinearLayout
    android:id="@+id/time_layout"
    android:layout_width="match_parent"
    android:layout_height="match_parent"
    android:orientation="vertical" >
<TextView
    android:id="@+id/tv_time"
    android:layout_width="wrap_content"
    android:layout_height="wrap_content"
    android:textSize="30sp"
    android:textColor="@android:color/black"
    android:layout_gravity="center_horizontal"/>
<TimePicker
    android:id="@+id/show_time"
    android:layout_width="wrap_content"
    android:layout_height="wrap_content"
    />
</LinearLayout>
<com.amap.api.maps.MapView
    android:id="@+id/map"
    android:layout_width="match_parent"
    android:layout_height="match_parent" />
</FrameLayout>
```

(5) 实现 MainActivity 中的功能

选项卡标签需要使用 LayoutInflater 类的 inflate(int resource, ViewGroup root, boolean attachToRoot) 方法完成实例化操作, 其中 3 个参数分别用来设置所需要的布局管理器的资源 ID、组件的容器和是否包含设置组件。本例中使用 activity_main 布局实例化选项卡标签。在添加标签时，有3种方法，参考代码中将逐一列出。

在加载地图时，需要用到MapView对象和AMap对象，二者的关系是：MapView是地图的容器，AMap对象用来显示地图。在使用MapView类时，必须要重写Activity类生命周期的所有方法，包括onCreate()、onDestroy()、onResume()、onPause()和onSaveInstanceState()。

参考代码如下：

```java
//使Activity类继承于TabActivity,以便生成选项卡
public class MainActivity extends TabActivity{
    //声明MapView对象
    //MapView是View类的一个子类,可以帮助开发者在View中放置地图
    //MapView是地图的容器,但通过AMap对象显示地图
    private MapView mapView;
    //声明AMap对象
    //AMap类将自动连接到高德地图服务,下载地图数据,在设备上显示地图等
    private AMap aMap;
    @Override
    public void onCreate(BundlesavedInstanceState){
        super.onCreate(savedInstanceState);
        //定义选项卡标签tabHost
        TabHost tabHost = this.getTabHost();
        //使用LayoutInflater类完成实例化操作
        //用activity_main布局实例化选项卡标签
        LayoutInflater.from(this).inflate(
            R.layout.activity_main,tabHost.getTabContentView(),true);
        //初始化日期显示文本和时间显示文本
        final TextView tvDate = (TextView)findViewById(R.id.tv_date);
        final DatePicker dateShow = (DatePicker)findViewById(R.id.show_date);
        tvDate.setText("今天是" + dateShow.getYear() + "年"
            + (dateShow.getMonth() + 1) + "月"
            + dateShow.getDayOfMonth() + "日");
        final TextView tvTime = (TextView)findViewById(R.id.tv_time);
        final TimePicker timeShow = (TimePicker)findViewById(R.id.show_time);
        tvTime.setText("现在是" + timeShow.getCurrentHour() + "时"
            + timeShow.getCurrentMinute() + "分");
    //3种增加Tab的写法
        //第1种写法
        TabSpec tabDate = tabHost.newTabSpec("date");
        tabDate.setIndicator("",this.getResources().getDrawable(R.drawable.images));
        tabDate.setContent(R.id.date_layout);
        tabHost.addTab(tabDate);
        //第2种写法
        TabSpec tabTime = tabHost.newTabSpec("time")
    .setIndicator("",this.getResources().getDrawable(R.drawable.clock))
    .setContent(R.id.time_layout);
        tabHost.addTab(tabTime);

        //关于地图的代码
```

```java
        mapView = (MapView)findViewById(R.id.map);
        mapView.onCreate(savedInstanceState);
        //完成地图初始化操作
        init();
        //第3种写法
        tabHost.addTab(tabHost.newTabSpec("map")
            .setIndicator("",this.getResources().getDrawable(R.drawable.map))
            .setContent(mapView.getId()));
}
/*** 初始化AMap对象 */
private void init(){
    if(aMap == null){
        aMap = mapView.getMap();
    }
}
/*** 方法必须重写 */
@Override
protected void onSaveInstanceState(Bundle outState){
    super.onSaveInstanceState(outState);
    mapView.onSaveInstanceState(outState);
}
/*** 此方法需要有 */
@Override
protected void onResume()
{
    super.onResume();
    mapView.onResume();
}
/*** 此方法需要有 */
@Override
protected void onPause(){
    super.onPause();
    mapView.onPause();
}
/*** 此方法需要有 */
@Override
protected void onDestroy(){
    super.onDestroy();
    mapView.onDestroy();
}}
```

4.2.2 选项卡（TabHost）的生成与使用

1. TabHost 和 TabActivity 的定义

选项卡 TabHost 的主要功能是对应用程序的各个模块进行分类管理。其主要特点是可以在一个窗口中显示多组标签的内容。在 Android 系统中，每个标签称为一个 Tab，而包含多个标签的容器就称为 TabHost。此类定义如下：

```
java. lang. Object
  ↳ android. view. View
    ↳ android. view. ViewGroup
      ↳ android. widget. FrameLayout
        ↳ android. widget. TabHost
```

在使用 TabHost 组件时，可以直接让自定义的 Activity 类继承 TabActivity 类。此类的定义如下：

```
java. lang. Object
  ↳ android. content. Context
    ↳ android. content. ContextWrapper
      ↳ android. view. ContextThemeWrapper
        ↳ android. app. Activity
          ↳ android. app. ActivityGroup
            ↳ android. app. TabActivity
```

2. TabHost 和 TabActivity 的常用方法

TabHost 类中定义的常用方法如表 4-6 所示。

表 4-6　TabHost 类的常用方法说明

序号	方法名称	说明
1	public FrameLayout getTabContentView( )	取得选项卡视图
2	public void addTab(TabHost. TabSpec tabSpec)	增加一个选项卡标签 Tab
3	public TabHost. TabSpec newTabSpec(String tag)	创建一个 TabSpec 对象
4	public void setup( )	如果 TabHost 对象是用 findViewById 的方法初始化，则在 addTab 前调用此方法，以建立 TabHost 对象 在 TabActivity 中，如果 TabHost 对象是使用 getTabHost 完成的初始化，则不需要再次调用此方法

TabActivity 类中定义的常用方法如表 4-7 所示。

表 4-7　TabActivity 类的常用方法说明

序号	方法名称	说明
1	public TabHost getTabHost( )	取得 TabHost 类的对象

3. LayoutInflater 的定义

在标签界面显示时，由于没有直接通过 TabHost 在 XML 中定义组件样式，所以无法使用 findViewById( ) 方法进行 TabHost 对象的实例化。此时，可以通过 LayoutInflater 完成布局管理器中定义组件的实例化操作。

此类定义如下：

```
java. lang. Object
  ↳ android. view. LayoutInflater
```

4. LayoutInflater 的常用方法

LayoutInflater 类的常用方法如表 4-8 所示。

表 4-8 LayoutInflater 类的常用方法说明

序号	方法名称	说明
1	public static LayoutInflater from(Context context)	从指定的环境中获得 LayoutInflater 对象
2	public View inflate(int resource, ViewGroup root, boolean attachToRoot)	设置所需要的布局管理器的资源 ID、组件的容器以及是否包含设置组件的参数

5. 取得 TabHost 对象并实例化

如前所述,调用 TabActivity 类的 getTabHost() 方法即可取得一个 TabHost 类的对象。本例中,使用如下代码取得 TabHost 类的对象 tabHost:

```
TabHost tabHost = this.getTabHost();
```

本例中,使用如下代码完成了从 activity_main.xml 布局管理器定义组件到 MainActivity.java 程序的实例化工作。

```
LayoutInflater.from(this).inflate(
    R.layout.activity_main,
    tabHost.getTabContentView(),
    true);
```

4.2.3 标签(TabSpec)的定义与使用

1. TabSpec 的定义

TabSpec 对象可以定义各个标签的标题、显示内容以及标签图标。此类定义如下:

```
java.lang.Object
  ↳ android.widget.TabHost.TabSpec
```

2. TabSpec 的常用方法

TabSpec 类的常用方法如表 4-9 所示。

表 4-9 TabSpec 类的常用方法说明

序号	方法名称	说明
1	public TabHost.TabSpec setContent(int viewId)	设置要显示的组件 ID
2	public TabHost.TabSpec setIndicator(CharSequence label)	设置标签的提示文本
3	public TabHost.TabSpec setIndicator(CharSequence label, Drawable icon)	设置标签的提示文本和显示图标

3. 增加 Tab 标签

增加标签 Tab 时,需要调用选项卡 TabHost 类的 addTab(TabHost.TabSpec tabSpec) 方法。本例中,分别使用了 3 种不同的方式增加并设置了 Tab 标签的各种常用属性。

第 1 种方式,先调用 newTabSpec(String tag) 方法创建一个 TabSpec 对象,然后再逐项为其设置常用属性,最后将新增加的 TabSpec 对象添加到选项卡 TabHost 中。参考代码如下:

```
TabSpec tabDate = tabHost.newTabSpec("date");
tabDate.setIndicator("",this.getResources().getDrawable(R.drawable.images));
tabDate.setContent(R.id.date_layout);
tabHost.addTab(tabDate);
```

第 2 种方式，在创建 TabSpec 对象的同时设置属性，然后添加到选项卡。参考代码如下：

```
TabSpec tabTime = tabHost.newTabSpec("time")
    .setIndicator("",this.getResources().getDrawable(R.drawable.clock))
        .setContent(R.id.time_layout);
tabHost.addTab(tabTime);
```

第 3 种方式，将代码进一步整合，直接添加标签，在添加的过程中才创建标签并为其设置属性。此时代码的可读性较差，不推荐使用。参考代码如下：

```
tabHost.addTab(
        tabHost.newTabSpec("map")
    .setIndicator("",this.getResources().getDrawable(R.drawable.map))
    .setContent(mapView.getId())
);
```

4.2.4 日期选择器（DatePicker）与时间选择器（TimePicker）

1. DatePicker 的定义

在 Android 中可以使用日期选择器 DatePicker 完成年、月、日的设置。此类定义如下：

```
java.lang.Object
    ↳ android.view.View
    ↳ android.view.ViewGroup
    ↳ android.widget.FrameLayout
    ↳ android.widget.DatePicker
```

在使用这个组件时，首先在 res/layout 中的 XML 文件里进行注册，注册时可参考如下代码：

```
<DatePicker
    android:id="@+id/show_date"
    android:layout_width="wrap_content"
    android:layout_height="wrap_content"/>
```

2. DatePicker 的常用方法

DatePicker 类的常用方法如表 4-10 所示。

表 4-10　DatePicker 类的常用方法说明

序号	方法名称	说　明
1	public int getDayOfMonth()	取得日期选择器所设置的日
2	public int getMonth()	取得日期选择器所设置的月
3	public int getYear()	取得日期选择器所设置的年
4	public void setEnabled(boolean enabled)	设置组件是否可用
5	public void init(int year, int monthOfYear, int dayOfMonth, DatePicker.OnDateChangedListener onDateChangedListener)	初始化日期并设置监听

3. TimePicker 的定义

与日期选择器相似，时间选择器 TimePicker 可以设置时间。此类定义如下：

```
java.lang.Object
    ↳ android.view.View
        ↳ android.view.ViewGroup
            ↳ android.widget.FrameLayout
                ↳ android.widget.TimePicker
```

在使用这个组件时,首先在 res/layout 中的 XML 文件里进行注册,注册时可参考如下代码:

```
<TimePicker
    android:id = "@+id/show_time"
    android:layout_width = "wrap_content"
    android:layout_height = "wrap_content"/>
```

4. TimePicker 的常用方法

TimePicker 类的常用方法如表 4-11 所示。

表 4-11 TimePicker 类的常用方法说明

序号	方法名称	说明
1	public Integer getCurrentHour()	取得当前时间选择器所设置的小时
2	public Integer getCurrentMinute()	取得当前时间选择器所设置的分钟
3	public boolean is24HourView()	判断是否是 24 小时制
4	public void setCurrentHour(Integer currentHour)	设置时间选择器的小时
5	public void setCurrentMinute(Integer currentMinute)	设置时间选择器的分钟
6	public void setIs24HourView(Boolean is24HourView)	设置时间是否为 24 小时制
7	public void setEnabled(boolean enabled)	设置时间选择器是否可用
8	public void setOnTimeChangedListener(TimePicker.OnTimeChangedListener onTimeChangedListener)	当时间修改时触发此事件

4.2.5 显示地图的基本步骤

1. 在布局文件中注册 MapView

本例使用高德地图 API,所以在注册时需要声明 MapView 类来源于 com.amap.api.maps.MapView,常用属性的设置方法与其他组件相似。

参考代码如下:

```
<com.amap.api.maps.MapView
    android:id = "@+id/map"
    android:layout_width = "match_parent"
    android:layout_height = "match_parent" />
```

2. 初始化 AMap 对象

前文已提,MapView 对象和 AMap 对象的关系是:MapView 是地图的容器,AMap 对象则用来显示地图。在程序运行时,如果地图值为空,则需要初始化。

参考代码如下:

```
        if( aMap == null) {
            aMap = mapView. getMap( );
        }
```

3. 重写 Activity 类生命周期的所有方法

特别提醒开发者，在使用 MapView 类时，必须要重写 Activity 类生命周期的所有方法。参考代码如下：

```
@Override
protected void onSaveInstanceState( Bundle outState) {
    super. onSaveInstanceState( outState) ;
    mapView. onSaveInstanceState( outState) ;
}
@Override
protected void onResume( )
{
    super. onResume( ) ;
    mapView. onResume( ) ;
}
@Override
protected void onPause( ) {
    super. onPause( ) ;
    mapView. onPause( ) ;
}
@Override
protected void onDestroy( ) {
    super. onDestroy( ) ;
    mapView. onDestroy( ) ;
}
```

4. 在 AndroidManifest. xml 中添加权限和密钥

在开发地图应用时，需要添加一系列权限，如可以访问网络，可以取得手机状态等。由于使用了第三方 API，需要得到第三方许可，所以还需要取得开发密钥。关于如何取得高德地图开发密钥，以及开发地图应用时需要添加哪些具体权限，前文已述，不再说明。

4.2.6　TabActivity 的取代者 FragmentActivity

在 3.0 以上版本的 Android 系统中，TabActivity 不再推荐使用，它被 FragmentActivity 取代。为了确定没有在旧版本系统上使用新版本的 APIs，应使用 android-support-v4. jar 包，因为它提供了 Fragment 的 APIs，使得在 Android 1.6（API level 4）以上的系统都可以使用 Fragment。

在 3.0 之前的项目中引用此包时，需要将包先拷入项目中的 libs 文件夹，然后在项目的 Properties 中添加。操作步骤为：选中该项目，在右键菜单中选中"Build Path"命令，在弹出的子菜单中单击"Configure Build Path"命令。然后在弹出的窗口中，单击"Add JARs…"按钮，在生成的选项卡中，选择项目 libs 下的"android-support-v4. jar"。单击"Order and Export"选项卡，勾选"android-support-v4. jar"选项，单击"OK"按钮完成包的添加。

1. FragmentActivity 的定义

当创建包含 Fragment 的 Activity 时，继承的应该是 FragmentActivity 而不是 Activity。此类

定义如下:

> java. lang. Object
> ↳ android. content. Context
> ↳ android. content. ContextWrapper
> ↳ android. view. ContextThemeWrapper
> ↳ android. app. Activity
> ↳ android. support. v4. app. FragmentActivity

2. FragmentActivity 的使用举例

例如,需要完成如图 4-8 所示的应用,除使用上文介绍的 TabActivity 之外,高版本中更推荐使用 FragmentActivity 实现。

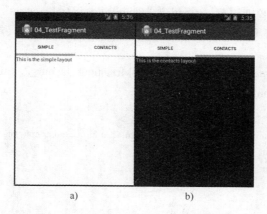

图 4-8 FragmentActivity 案例示意图

3. FragmentActivity 的使用技巧

1) 在 res/layout 中定义选项卡布局 activity_main.xml。

参考代码如下:

```
<?xml version = "1.0" encoding = "utf-8"?>
<android. support. v4. app. FragmentTabHost
    xmlns:android = "http://schemas. android. com/apk/res/android"
android:id = "@android:id/tabhost"
    android:layout_width = "match_parent"
    android:layout_height = "match_parent" >
<LinearLayout
    android:orientation = "vertical"
    android:layout_width = "match_parent"
    android:layout_height = "match_parent" >
        <TabWidget
        android:id = "@android:id/tabs"
        android:orientation = "horizontal"
        android:layout_width = "match_parent"
        android:layout_height = "wrap_content"
        android:layout_weight = "0"/>
<FrameLayout
```

```xml
            android:id = "@android:id/tabcontent"
            android:layout_width = "0dp"
            android:layout_height = "0dp"
            android:layout_weight = "0"/>
    <FrameLayout
            android:id = "@+id/realtabcontent"
            android:layout_width = "match_parent"
            android:layout_height = "0dp"
            android:layout_weight = "1"/>
</LinearLayout>
</android.support.v4.app.FragmentTabHost>
```

注意：在此布局中必须得有一个 id 为"@android:id/tabcontent"的 FrameLayout 和一个 id 为"@android:id/tabs"的 TabWidget。关于 TabWidget 的定义，感兴趣的读者可以自行到 Android 开发者文档中查阅。

2）定义标签布局 fragment_layout_1.xml 和 fragment_layout_2.xml。

fragment_layout_1.xml 的参考代码如下：

```xml
<?xml version = "1.0" encoding = "utf-8"?>
<LinearLayout xmlns:android = "http://schemas.android.com/apk/res/android"
    android:layout_width = "match_parent"
    android:layout_height = "match_parent"
    android:orientation = "vertical" >
<TextView
        android:id = "@+id/tv_left"
        android:layout_width = "wrap_content"
        android:layout_height = "wrap_content"
        android:text = "This is the simple layout"
        android:textColor = "@android:color/black"/>
</LinearLayout>
```

fragment_layout_2.xml 的参考代码如下：

```xml
<?xml version = "1.0" encoding = "utf-8"?>
<LinearLayout xmlns:android = "http://schemas.android.com/apk/res/android"
    android:layout_width = "match_parent"
    android:layout_height = "match_parent"
    android:orientation = "vertical"
    android:background = "@android:color/black" >
<TextView
        android:id = "@+id/tv_right"
        android:layout_width = "wrap_content"
        android:layout_height = "wrap_content"
        android:text = "This is the contacts layout"
        android:textColor = "@android:color/darker_gray"/>
</LinearLayout>
```

3）创建各个标签对应的 Fragment 子类。

fragment_layout_1.xml 对应的 Fragment 子类 FragmentA1.java 的参考代码如下：

```
package com. book. testfragment;

import android. os. Bundle;
import android. support. v4. app. Fragment;
import android. view. LayoutInflater;
import android. view. View;
import android. view. ViewGroup;

public class FragmentA1 extends Fragment {
    @Override
    public ViewonCreateView(LayoutInflater inflater, ViewGroup container,
            Bundle savedInstanceState) {
        return inflater. inflate(R. layout. fragment_layout_1, container, false);
    }
}
```

fragment_layout_2. xml 对应的 Fragment 子类 FragmentA2. java 的参考代码如下:

```
package com. book. testfragment;

import android. os. Bundle;
import android. support. v4. app. Fragment;
import android. view. LayoutInflater;
import android. view. View;
import android. view. ViewGroup;

public class FragmentA2 extends Fragment {
    @Override
    public ViewonCreateView(LayoutInflater inflater, ViewGroup container,
            Bundle savedInstanceState) {
        // TODO Auto - generated method stub
        return inflater. inflate(R. layout. fragment_layout_2, container, false);
    }
}
```

注意：这两个类都继承于 Fragment，在使用时需要引入 android. support. v4. app. Fragment 包。

4) 创建 FragmentActivity 的子类，添加标签。

```
public class MainActivity extends FragmentActivity {
    private FragmentTabHost mTabHost;
    @Override
    protected void onCreate(BundlesavedInstanceState) {
        super. onCreate(savedInstanceState);
        //设置当前类对应的 layout
        setContentView(R. layout. activity_main);
        //初始化 TabHost 对象
        mTabHost = (FragmentTabHost) findViewById(android. R. id. tabhost);
        //在 addTab 之前创建 TabHost 对象
        mTabHost. setup(this, getSupportFragmentManager(), R. id. realtabcontent);
        //添加标签
        mTabHost. addTab(mTabHost. newTabSpec("simple"). setIndicator("Simple"),
                FragmentA1. class, null);
```

```
            mTabHost.addTab(mTabHost.newTabSpec("contacts").setIndicator("Contacts"),
                    FragmentA2.class,null);
        }
    }
```

4.3 实例3：全球名校快查

4.3.1 功能要求与操作步骤

1. 功能要求

完成如图4-9所示的应用程序。要求能够：

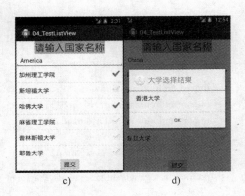

图4-9 "全球名校快查"运行效果图

① 在图4-9a所示的界面中，在自动完成文本框中输入国家名称首字母，即可出现与此首字母相同的国家名称列表。

② 在用户从列表中选择了国家名称之后，即可弹出如图4-9b所示的界面，显示该国的名校名称。各个列表项可以被选择，选择效果界面如图4-9c所示。

③ 当用户选择了某些列表项并单击下方的"提交"按钮后，可以弹出选择结果，效果如图4-9d所示。

2. 操作步骤

（1）创建项目

参考如下输入信息创建 Android 项目：

  Application Name：04_TestListView
  Project Name：04_TestListView
  Package Name：com.book.testlistview
  Activity Name：MainActivity
  Layout Name：activity_main

（2）定义布局

从图4-9各图可看出，本应用可以使用垂直线性布局或者相对布局来实现，此处以垂直线性布局为例来定义布局和组件。可参考表4-12设置各个组件的属性。

表 4-12　实例 3 中各个组件属性列表

序号	组件类型	属性名称	属性值	说明
1	LinearLayout	android:layout_width	match_parent	宽度匹配父容器
		android:layout_height	match_parent	高度匹配父容器
		android:orientation	vertical	垂直方向
2	TextView	android:layout_width	wrap_content	宽度适合内容
		android:layout_height	0dp	高度与权重配合使用
		android:layout_weight	1	高度占布局的 10%
		android:text	请输入国家名称	设置显示文本
		android:textSize	30sp	字号为 30sp
		android:layout_gravity	center	文本框居中显示
		android:background	@android:color/darker_gray	设置文本框背景颜色
3	AutoComplete-TextView	android:id	@+id/actv_country	添加 id 为 "actv_country" 的组件
		android:layout_width	match_parent	宽度匹配父容器
		android:layout_height	0dp	高度与权重配合使用
		android:layout_weight	1	高度占布局的 10%
		android:completionThreshold	1	设置用户输入 1 个字符后即给出提示
		android:ems	10	设置文本组件的宽度为 10 个字符的宽度
4	ListView	android:id	@+id/lv_universities	为组件设置 id
		android:layout_width	match_parent	宽度匹配父容器
		android:layout_height	0dp	高度与权重配合使用
		android:layout_weight	7	高度占布局的 70%
5	Button	android:layout_gravity	center	将组件居中显示
		android:layout_width	wrap_content	宽度适合内容即可
		android:layout_height	0dp	高度与权重配合使用
		android:layout_weight	1	高度占布局的 10%
		android:text	提交	组件显示文本
		android:onClick	clickSubmit	单击事件

参考代码如下：

```
<LinearLayout xmlns:android="http://schemas.android.com/apk/res/android"
    xmlns:tools="http://schemas.android.com/tools"
    android:layout_width="match_parent"
    android:layout_height="match_parent"
    android:orientation="vertical" >
    <TextView
        android:layout_width="wrap_content"
        android:layout_height="0dp"
        android:layout_weight="1"
        android:text="请输入国家名称"
        android:textSize="30sp"
        android:layout_gravity="center"
        android:background="@android:color/darker_gray"/>
    <AutoCompleteTextView
        android:id="@+id/actv_country"
        android:layout_width="match_parent"
```

```xml
            android:layout_height = "0dp"
            android:layout_weight = "1"
            android:ems = "10"
            android:completionThreshold = "1" >
    <requestFocus />
    </AutoCompleteTextView>
    <ListView
            android:id = "@+id/lv_universities"
            android:layout_width = "match_parent"
            android:layout_height = "0dp"
            android:layout_weight = "7"  >
    </ListView>
    <Button
            android:layout_gravity = "center"
            android:layout_width = "wrap_content"
            android:layout_height = "0dp"
            android:layout_weight = "1"
            android:text = "提交"
            android:onClick = "clickSubmit" />
</LinearLayout>
```

(3) 实现 MainActivity 中的功能

为了能够使自动完成文本框根据用户输入的字符给出提示列表项，需要预先定义常量字符串数组（本例中，字符串数组名称为 COUNTRIES），然后为其设置适配器。同理，为了使列表视图可以根据自动完成文本框所选项，给出相关列表项，也需要预先定义常量字符串数组（本例中，字符串数组名称为 UNIVERSITIES），然后再为其设置适配器。UNIVERSITIES 数组需为二维数组，行表示与 COUNTRIES 数组项对应的国家，列表示该国家所包含的名校。

参考代码如下：

```java
public class MainActivity extends Activity implements  OnItemClickListener{
    //两个常量字符串数组分别用来存放国家名称和对应的大学名称
    private static final String[] COUNTRIES = new String[] {
        "America",
        "Austrialia",
        "China",
        "Canada",
        "England",
        "France",
        "Germany",
        "Japan",
        "Sinapore"};
    private static final String[][] UNIVERSITIES = new String[][] {
        {"加州理工学院","斯坦福大学","哈佛大学","麻省理工学院","普林斯顿大学","耶鲁大学"},
        {"墨尔本大学","澳大利亚国立大学","悉尼大学"},
        {"香港大学","北京大学","清华大学","香港科技大学","复旦大学"},
        {"多伦多大学"},
        {"牛津大学","剑桥大学","伦敦帝国理工学院","伦敦大学学院","英属哥伦比亚大学","爱丁堡大学"},
```

```java
                    {"巴黎第六大学","巴黎第十一大学"},          {"慕尼黑大学","海德堡大学"},
                    {"东京大学","京都大学","名古屋大学"},
                    {"新加坡国立大学","南洋理工大学"}};
    //声明组件
    ListView lvUniversities;
    AutoCompleteTextView actvCountry;
    //字符串数组 universityName 用来临时存储选中国家后对应的大学名称

    String[] universityName = null;
    //字符串用来存储用户选中的大学名称

    String universityChosen = "";
    @Override
    protected void onCreate(Bundle savedInstanceState) {
        super.onCreate(savedInstanceState);
        setContentView(R.layout.activity_main);
        //为自动完成文本框定义适配器
        ArrayAdapter<String> actvAdapter = new ArrayAdapter<String>(
            this,
            android.R.layout.simple_dropdown_item_1line,
            COUNTRIES);
        //定义布局中的组件
        lvUniversities = (ListView) findViewById(R.id.lv_universities);
        actvCountry = (AutoCompleteTextView) findViewById(R.id.actv_country);
        //设置适配器
        actvCountry.setAdapter(actvAdapter);
        //设置单击监听事件
        actvCountry.setOnItemClickListener(this);
    }
    //实现自动完成文本框的单击监听事件
    @Override
    public void onItemClick(AdapterView<?> arg0,View v,int position,long id) {
            //根据自动完成文本框中的文本,确定列表视图中的列表项
            if(actvCountry.getText().toString().equals(COUNTRIES[0])){
                universityName = UNIVERSITIES[0];
            }else if(actvCountry.getText().toString().equals(COUNTRIES[1])){
                universityName = UNIVERSITIES[1];
            }else if(actvCountry.getText().toString().equals(COUNTRIES[2])){
                universityName = UNIVERSITIES[2];
            }else if(actvCountry.getText().toString().equals(COUNTRIES[3])){
                universityName = UNIVERSITIES[3];
            }else if(actvCountry.getText().toString().equals(COUNTRIES[4])){
                universityName = UNIVERSITIES[4];
            }else if(actvCountry.getText().toString().equals(COUNTRIES[5])){
                universityName = UNIVERSITIES[5];
            }else if(actvCountry.getText().toString().equals(COUNTRIES[6])){
                universityName = UNIVERSITIES[6];
            }else if(actvCountry.getText().toString().equals(COUNTRIES[7])){
                universityName = UNIVERSITIES[7];
```

```
        }else if(actvCountry.getText().toString().equals(COUNTRIES[8])){
            universityName = UNIVERSITIES[8];
        }
        //如果自动完成文本框中有输入项,则为列表视图设置适配器
        if(!actvCountry.getText().equals(null)){
            //定义适配器
            ArrayAdapter<String> lvAdapter = new ArrayAdapter<String>(
                    this,
                    android.R.layout.simple_list_item_checked,
                    universityName);
            //设置适配器
            lvUniversities.setAdapter(lvAdapter);
            //设置列表视图的选择模式
            lvUniversities.setChoiceMode(AbsListView.CHOICE_MODE_MULTIPLE);
            //实现单击事件
            OnItemClickListener clickLV = new OnItemClickListener(){
                @Override
                public void onItemClick(AdapterView<?> arg0, View v,
                        int position, long id){
                    //将选择的列表文本连接后存放到字符串 universityChosen 中

                    universityChosen = universityChosen + " " + universityName[position];

                }
            };
            //为列表视图添加单击事件
            lvUniversities.setOnItemClickListener(clickLV);
        }else{
            Toast.makeText(MainActivity.this,"请输入或选择国家名称!",
                    Toast.LENGTH_SHORT).show();
        }
    }
    public void clickSubmit(View v){
        //构造对话框,并显示
        AlertDialog.Builder resultDialog = new AlertDialog.Builder(MainActivity.this);
        resultDialog.setTitle("大学选择结果")
                .setIcon(android.R.drawable.ic_dialog_info)
                .setMessage(universityChosen)
                .setPositiveButton("OK",null)
                .create()
                .show();
    }
}
```

4.3.2 自动完成文本框(AutoCompleteTextView)的定义与使用

1. AutoCompleteTextView 的定义

自动完成文本框 AutoCompleteTextView 的主要功能是在用户输入一定数量字符后,该组件根据适配器内容提供一个下拉列表,供用户从中选择,当用户选择某个列表项之后,组件

按照选择自动填写文本框。此类定义如下：

> java. lang. Object
> ↳ android. view. View
> ↳ android. widget. TextView
> ↳ android. widget. EditText
> ↳ android. widget. AutoCompleteTextView

在使用这个组件时，首先在 res/layout 中的 XML 文件里进行注册，注册时可参考如下代码：

```
<AutoCompleteTextView
    android:id = "@+id/actv_country"
    android:layout_width = "match_parent"
    android:layout_height = "0dp"
    android:layout_weight = "1"
    android:ems = "10"
    android:completionThreshold = "1" >
</AutoCompleteTextView>
```

在 XML 文件中，可以使用 android:completionThreshold 属性设置用户需要输入几个字符即可给出提出下拉列表框提示。从该类的定义可以看出，是从 EditText 继承而来，所以它实际上也是一个文本编辑框，只是多了一个自动提示输入补全的功能。

2. AutoCompleteTextView 的常用方法

从该类的定义中可以看出，AutoCompleteTextView 类是 TextView 类的子类，所以其常用方法中有一些继承于 TextView，如表 4-13 所示。

表 4-13 AutoCompleteTextView 类的常用方法说明

序号	方法名称	说明
1	public void setAdapter (T adapter)	设置适配器
2	public void setOnItemClickListener (AdapterView. OnItemClickListener l)	设置列表项单击事件
3	public Editable getText( )	继承于 EditText 类的方法，获取自动完成文本框的文本内容

3. AutoCompleteTextView 的使用步骤

在使用自动完成文本框时，需要首先定义列表项需要使用的适配器。参考代码如下：

```
ArrayAdapter<String> actvAdapter = new ArrayAdapter<String>(
        this,
        android.R.layout.simple_dropdown_item_1line,
        COUNTRIES);
```

在定义 actvAdapter 适配器时，需要有 3 个参数，分别是当前环境变量 this、适配器使用的布局类型 android.R.layout.simple_dropdown_item_1line 以及在目标列表中显示的元素，此处为字符串数组 COUNTRIES。

然后，自动完成文本框对象（本例为 actvCountry）调用 setAdapter(T adapter)方法使用适配器。参考代码如下：

```
actvCountry.setAdapter(actvAdapter);
```

最后，为了使自动完成文本框能够响应单击事件，需要为其设置单击监听。参考代码如下：

```
actvCountry.setOnItemClickListener(this);
```

使用上述代码添加监听时，需要让 this 所代表的当前类实现 OnItemClickListener 类的 onItemClick(AdapterView<?> arg0, View v, int position, long id) 方法。

4.3.3 列表视图（ListView）的定义与使用

1. ListView 的定义

列表视图 ListView 的主要功能是以垂直列表的形式列出需要显示的列表项。此类定义如下：

```
java.lang.Object
 ↳ android.view.View
   ↳ android.view.ViewGroup
     ↳ android.widget.AdapterView<T extends android.widget.Adapter>
       ↳ android.widget.AbsListView
         ↳ android.widget.ListView
```

在使用这个组件时，有两种方式可以实现，一种是在 res/layout 中的 XML 文件里进行注册，然后使用；另一种是让 Activity 继承 ListActivity 实现。实例 3 中使用了第 1 种方式。在 XML 文件中注册时可参考如下代码：

```xml
<ListView
    android:id="@+id/lv_universities"
    android:layout_width="match_parent"
    android:layout_height="0dp"
    android:layout_weight="7">
</ListView>
```

在 xml 文件中，可以使用 android:entries 属性设置列表项需要显示的数组资源。例如，android:entries="@array/countries"。使用此属性时，则需要在 res/values 目录中创建数组资源文件 arrays.xml，并在该文件中添加名称为"countries"的字符串数组，参考代码如下：

```xml
<resources>
    <string-array name="countries">
        <item>中国</item>
        <item>美国</item>
        <item>英国</item>
        <item>日本</item>
        <item>澳大利亚</item>
    </string-array>
</resources>
```

此时在定义适配器时，需要调用 ArrayAdapter 类的 createFromResource(Context c, int textResourceId,int textViewResourceId）方法来创建 ArrayAdapter 对象。参考代码如下：

```
ArrayAdapter < CharSequence > lvAdapter = ArrayAdapter.createFromResource(
                    this,
                    R.array.countries,
                    android.R.layout.simple_dropdown_item_1line);
```

2. ListView 的常用方法

从类的定义中可以看出，ListView 类是 View 类的子类，所以其常用方法中有一些继承于 View，如表 4-14 所示。

表 4-14 ListView 类的常用方法说明

序号	方法名称	说明
1	public void setAdapter（ListAdapter adapter）	设置适配器
2	public void setChoiceMode（int choiceMode）	继承于 AbsListView 的方法，用来设置选择模式，可选值有 CHOICE_MODE_MULTIPLE、CHOICE_MODE_MULTIPLE_MODAL、CHOICE_MODE_SINGLE 和 CHOICE_MODE_NONE
3	public void setOnItemClickListener（AdapterView.OnItemClickListener listener）	为列表项设置单击监听

3. ListView 的使用技巧

与自动完成文本框相似，在使用 ListView 时，需要首先定义列表项需要使用的适配器。参考代码不再赘述。然后，ListView 对象调用 setAdapter(T adapter) 方法使用适配器。最后，为 ListView 设置单击监听，并实现监听事件。

4.4 动手实践 4：休闲时分

4.4.1 功能要求

1. 应用程序界面

应用程序的界面如图 4-10 所示。

图 4-10 "休闲时分"运行效果图

2. 功能描述

用户单击图4-10a"笑话连篇"中的任何一个笑话标题后，会弹出如图4-10b的对话框，对话框的标题中显示笑话标题和信息图标，对话框的主体部分显示笑话的内容，下方有一个"OK"按钮。

单击图4-10a中的"美图欣赏"后，可以随机出现一张图片，效果如图4-10c所示。

4.4.2 操作提示

首先将程序中需要的图片资源准备好，然后按照以下步骤完成应用程序。

1. 完成应用程序首页面

在设计 res/layout/activity_main.xml 时，为了有选项卡效果，需要用到 TabWidget 组件，而这个组件需要放在 TabHost 容器中。若想将选项卡放在页面下方，需要使用到一些技巧。如将 RadioGroup + RadioButton 组合使用，不将 TabWidget 显示在页面中等。这个页面布局比较复杂，可以参考以下代码设计。

```xml
<?xml version = "1.0" encoding = "utf-8"?>
<!--TabHost 的 id 必须为@android:id/tabhost -->
<TabHost
    android:id = "@android:id/tabhost"
    android:layout_width = "fill_parent"
    android:layout_height = "fill_parent"
    xmlns:android = "http://schemas.android.com/apk/res/android" >
<LinearLayout
        android:orientation = "vertical"
        android:layout_width = "fill_parent"
        android:layout_height = "fill_parent" >
<!--在 TabHost 中必须有一个 id 为@android:id/tabcontent 的 FrameLayout -->
<FrameLayout
        android:id = "@android:id/tabcontent"
        android:layout_width = "fill_parent"
        android:layout_height = "0.0dip"
        android:layout_weight = "1.0"/>
<!--TabWidget 的 id 必须为 @android:id/tabs -->
<!--TabWidget 的 visibility 值为 gone 时,表示隐藏组件,
        值为 invisible 时,表示不可见组件 -->
<!--值为 invisible 时界面保留了组件所占空间；
        若为 gone 则不保留组件占有空间 -->
<TabWidget
        android:id = "@android:id/tabs"
        android:visibility = "gone"
        android:layout_width = "fill_parent"
        android:layout_height = "wrap_content"
        android:layout_weight = "0.0"/>
<!--RadioGroup 底部对齐,两个子组件水平方向排列,并在垂直方向上居中对齐 -->
<RadioGroup
```

```xml
android:id = "@+id/rg_main"
android:gravity = "center_vertical"
android:layout_gravity = "bottom"
android:orientation = "horizontal"
android:paddingBottom = "2dip"
android:paddingTop = "8dp"
android:background = "@drawable/tab_back_color"
android:layout_width = "fill_parent"
android:layout_height = "wrap_content" >
<!-- 将第1个单选钮设置为默认被选中的状态 -->
<!-- 将单选钮的上方图片用 tab_btn_selector 设置,
     代码见 res/drawable/tab_btn_selector.xml -->
<!-- 将单选钮的提示文本颜色用 tab_text_selector 设置,
     代码见 res/drawable/tab_text_selector.xml -->
<!-- 将单选钮的样式用 MMTabButton 样式设置,
     代码见 res/values/styles.xml -->
<RadioButton
    android:id = "@+id/tab_joke"
    android:checked = "true"
    android:text = "@string/joke_text"
    android:drawableTop = "@drawable/tab_btn_selector"
    android:textColor = "@drawable/tab_text_selector"
    style = "@style/MMTabButton"/>
<RadioButton
    android:id = "@+id/tab_pic"
    android:text = "@string/pic_text"
    android:drawableTop = "@drawable/tab_btn_selector"
    android:textColor = "@drawable/tab_text_selector"
    style = "@style/MMTabButton"/>
</RadioGroup>
</LinearLayout>
</TabHost>
```

如果想要添加其他选项标签,则只需要在上述代码中增加 RadioButton 并对其属性进行设置即可。

2. 定义常量类

在常量类 Constant.java 中定义标签 Tag 字符串和两个标签对应的 Activity 页面类。参考代码如下:

```java
//常量类
public class Constant {
    //两个标签的 Tag
    public static String TabSpecesArray[ ] = {"笑话","美图"};
    //两个标签对应的 Activity 类
    public static Class <?> TabClassesArray[ ] = {
        JokeActivity.class,
        PicActivity.class,
```

        };
    }

3. 定义辅助样式

单选按钮其他样式 res/values/styles.xml：

```xml
<resources>
<style name="AppBaseTheme" parent="android:Theme.Light"/>
<style name="AppTheme" parent="AppBaseTheme"/>

<!--新增文本样式-->
<style name="tab_item_text_style">
<item name="android:textSize">12.0dip</item>
<item name="android:textColor">@drawable/tab_text_selector</item>
<item name="android:ellipsize">marquee</item>
<item name="android:singleLine">true</item>
</style>

<!--单选按钮其他样式-->
<style name="MMTabButton">
<item name="android:textAppearance">@style/tab_item_text_style</item>
<item name="android:gravity">center_horizontal</item>
<item name="android:layout_width">0.0dip</item>
<item name="android:layout_height">wrap_content</item>
<item name="android:button">@null</item>
<item name="android:layout_weight">1.0</item>
</style>
</resources>
```

单选按钮状态改变时图片使用的样式 res/drawable/tab_btn_selector.xml：

```xml
<?xml version="1.0" encoding="utf-8"?>
<selector xmlns:android="http://schemas.android.com/apk/res/android">
<item
        android:state_enabled="true"
        android:state_checked="true"
        android:drawable="@drawable/pressed_btn"/>
<item
        android:drawable="@drawable/initial_btn"/>
</selector>
```

单选按钮状态改变时文本的使用样式 res/drawable/tab_text_selector.xml：

```xml
<?xml version="1.0" encoding="utf-8"?>
<selector xmlns:android="http://schemas.android.com/apk/res/android">
<item
        android:state_checked="true"
        android:color="@android:color/white"/>
```

```
< item android:color = "#ff666666"/>
</selector>
```

4. 准备所需资源

字符串资源 res/values/strings.xml：

```
<?xml version = "1.0" encoding = "utf-8"?>
<resources>
<string name = "app_name">Ex04_1</string>
<string name = "action_settings">Settings</string>
    <string name = "joke_text">笑话连篇</string>
    <string name = "pic_text">美图欣赏</string>
</resources>
```

数组资源 res/values/arrays.xml：

```
<?xml version = "1.0" encoding = "utf-8"?>
<resources>
<!-- 定义各个笑话的题目 -->
<string-array name = "joke_name">
<item>真倒霉</item>
<item>快叫医生</item>
<item>穷秀才</item>
其他项省略...
</string-array>
<!-- 定义各个笑话的内容 -->
<string-array name = "joke_content">
<item>
一美女下班回家...
    </item>
<item>
约翰先生坐火车出远门..
    .</item>
<item>
有个人刚死,来见阎王...
    </item>
其他项省略...
</string-array>
</resources>
```

单击"笑话连篇"时的布局资源 res/layout/joke_layout.xml：

```
<?xml version = "1.0" encoding = "utf-8"?>
<LinearLayout xmlns:android = "http://schemas.android.com/apk/res/android"
    android:layout_width = "match_parent"
    android:layout_height = "match_parent"
    android:orientation = "vertical">
<!-- ListView 的 id 必须为@android:id/list -->
```

```xml
<!-- 用 entries 直接定义 ListView 的列表项 -->
<ListView
        android:id = "@android:id/list"
        android:layout_width = "match_parent"
        android:layout_height = "wrap_content"
        android:entries = "@array/joke_name" >
</ListView>
</LinearLayout>
```

单击"美图欣赏"时的布局资源 res/layout/pic_layout.xml：

```xml
<?xml version = "1.0" encoding = "utf-8"?>
<LinearLayout xmlns:android = "http://schemas.android.com/apk/res/android"
        android:layout_width = "match_parent"
        android:layout_height = "match_parent"
        android:orientation = "vertical" >
<ImageView
        android:id = "@+id/img_show"
        android:layout_width = "fill_parent"
        android:layout_height = "match_parent"
        android:src = "@drawable/background_pic"/>
</LinearLayout>
```

5. 实现 MainActivity 中的功能

参考代码如下：

```java
public class MainActivity extends TabActivity {
    //定义需要的组件
    private TabHost      tabHost;
    private RadioGroup   rgMain;
    public void onCreate(Bundle savedInstanceState) {
        super.onCreate(savedInstanceState);
        setContentView(R.layout.activity_main);
        //初始化
        init();
    }
    private void init()
    {
        //获取布局中的 TabHost
        tabHost = getTabHost();
        //从自定义的常量类 Constant 中取得标签页面类的长度
        int count = Constant.TabClassesArray.length;
        for(int i = 0; i < count; i++)
        {
            //创建标签,Tag 为常量类中的 TabSpecesArray 元素
            TabSpec tabSpec = tabHost.newTabSpec(Constant.TabSpecesArray[i])
                    .setIndicator(Constant.TabSpecesArray[i])
                    //将 Tab 中的内容设为一个用 getTabItemIntent(i)得到的意图
                    .setContent(getTabItemIntent(i));
```

```
                //添加标签
                tabHost.addTab(tabSpec);
        }
        //获取单选按钮组合组件
        rgMain = (RadioGroup) findViewById(R.id.rg_main);
        //为单选按钮组添加选择状态更改监听
        rgMain.setOnCheckedChangeListener(new OnCheckedChangeListener() {
            @Override
            public void onCheckedChanged(RadioGroup group, int checkedId) {
                //判断哪个选项卡标签被选中
                switch(checkedId) {
                    case R.id.tab_joke:
                        tabHost.setCurrentTabByTag(Constant.TabSpecesArray[0]);
                        break;
                    case R.id.tab_pic:
                        tabHost.setCurrentTabByTag(Constant.TabSpecesArray[1]);
                        break;
                }
            }
        });
        //改变单选按钮组中第1个单选按钮的选中状态
        //Change the checked state of the view to the inverse of its current state
        ((RadioButton) rgMain.getChildAt(0)).toggle();
    }
    //根据选择的标签决定启动哪个页面类
    private Intent getTabItemIntent(int index)
    {
        Intent intent = new Intent(this, Constant.TabClassesArray[index]);
        return intent;
    }
}
```

6. 定义 JokeActivity.java 类

在单击"笑话连篇"时，需要显示 JokeActivity 类，该类的完整定义如下：

```
public class JokeActivity extends ListActivity {
    @Override
    protected void onCreate(Bundle savedInstanceState) {
        super.onCreate(savedInstanceState);
        this.setContentView(R.layout.joke_layout);
        ArrayAdapter<CharSequence> jokeAdapter = ArrayAdapter.createFromResource(
            this,
            R.array.joke_name,
            android.R.layout.simple_list_item_1);
        this.setListAdapter(jokeAdapter);
    }
    @Override
    protected void onListItemClick(ListView l, View v, int position, long id) {
```

```
AlertDialog.Builder resultDialog = new AlertDialog.Builder(JokeActivity.this);
String[] jokeFromRes = getResources().getStringArray(R.array.joke_content);
CharSequence jokeContent = jokeFromRes[position];
resultDialog.setTitle(l.getItemAtPosition(position).toString() + "详情")
    .setIcon(android.R.drawable.ic_menu_info_details)
    .setMessage(jokeContent)
    .setPositiveButton("OK", null)
    .create()
    .show();
    }
}
```

7. 定义 PicActivity.java 类

在单击"美图欣赏"时，需要显示 PicActivity 类，该类的完整定义如下：

```
public class PicActivity extends Activity {
    @Override
    protected void onCreate(Bundle savedInstanceState) {
        super.onCreate(savedInstanceState);
        this.setContentView(R.layout.pic_layout);
    }
    //当 Activity"继续"显示时,执行 onResume 中的过程
    @Override
    protected void onResume() {
        super.onResume();
        int[] imgIDs = {R.drawable.boiled_mutton2,
            R.drawable.duck,
            R.drawable.peking_duck,
            R.drawable.peking_duck_2,
            R.drawable.send_message,
            R.drawable.tx04,
            R.drawable.tx06};
        ImageView imgShow = (ImageView) findViewById(R.id.img_show);
        Random randImgId = new Random();
        int i = randImgId.nextInt(imgIDs.length);
        imgShow.setImageResource(imgIDs[i]);
    }
}
```

8. 在 AndroidManifest.xml 中配置 Activity

```
<activity
    android:name="MainActivity"
    android:label="@string/app_name" >
<intent-filter>
<action android:name="android.intent.action.MAIN"/>
<category android:name="android.intent.category.LAUNCHER"/>
</intent-filter>
</activity>
```

```xml
< activity
    android:name = "JokeActivity"
    android:label = "@string/app_name" >
</activity>
< activity
    android:name = "PicActivity"
    android:label = "@string/app_name" >
</activity>
```

第5章 Android 应用程序的组成

完成动手实践4时运用了3个 Activity 实现界面切换。在有的应用程序中不仅需要多个 Activity，而且还需要在不同的 Activity 之间实现数据通信。此时，需要用到 Android 的其他组件。任何 Android 程序都可以由以下4部分组成，它们分别是 Activity，Broadcast Receiver，Service，Content Provider。使用时开发者需要在配置文件 AndroidManifest.xml 中声明项目中所有使用到的上述4种组件名称，如果有必要还需为每个组件的功能和需求进行相关描述。本章将对这4种组件进行详细介绍。

5.1 实例1：身体质量指数测试

由于个人的体型骨架不同，为了同时顾及身高和体重的配合，采用的指标是身体质量指数（Body Mass Index，BMI），它是与体内脂肪总量密切相关的指标，主要反映全身性超重和肥胖。由于 BMI 计算的是身体脂肪的比例，所以在测量身体因超重而面临心脏病、高血压等风险上，比单纯的以体重来认定更具准确性。该指数适用于所有18至65岁的人士，但儿童、发育中的青少年、孕妇、乳母、老人及身形健硕的运动员除外。其计算公式如下：

$$BMI = 体重(kg) / 身高^2(m^2)$$

健康 BMI 标准为：$18.5 \leq BMI < 24$。如果 $24 \leq BMI < 28$ 则超重，BMI 超过28则属于肥胖，此时不利于身体健康。世界卫生组织认为 BMI 指数保持在22左右是比较理想的。

5.1.1 功能要求与操作步骤

1. 功能要求

完成如图5-1所示的应用程序。要求能够：

① 允许用户输入浮点类型的身高值和体重值，如图5-1a 所示。

② 在用户输入数据后，单击"计算"按钮，切换到"身体质量指数"界面，该界面用

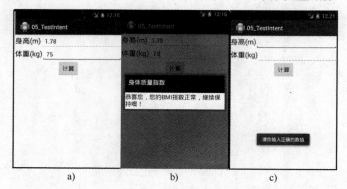

图5-1 "身体质量指数测试"运行效果图

对话框的形式显示，如图 5-1b 所示。

③ 如果输入的数值为空，则给出如图 5-1c 所示的提示。

2. 操作步骤

（1）创建项目

参考如下输入信息创建 Android 项目：

> Application Name：05_TestIntent
> Project Name：05_TestIntent
> Package Name：com. book. testintent
> Activity Name：MainActivity
> Layout Name：activity_main

（2）定义 activity_main.xml 布局

本布局需要用到的组件及各个组件的相关属性如表 5-1 所示。

表 5-1 "身体质量指数测试"中首界面属性列表

序号	组件类型	属性名称	属性值	说明
1	LinearLayout（根节点）	android:layout_width	match_parent	宽度匹配父容器
		android:layout_height	match_parent	高度匹配父容器
		android:orientation	vertical	垂直方向
2	LinearLayout（包含1个文本框和文本编辑框）	android:layout_width	match_parent	宽度匹配父容器
		android:layout_height	wrap_content	高度适合内容即可
		android:orientation	horizontal	水平方向
3	两个 TextView	android:id	@+id/tv_height @+id/tv_weight	添加 id 为 "tv_height" 和 "tv_weight" 的组件
		android:layout_width	wrap_content	宽度适合内容即可
		android:layout_height	wrap_content	高度适合内容即可
		android:text	身高(m)(或体重(kg))	提示文本
		android:textAppearance	?android:attr/textAppearanceLarge	设置大文本显示
4	两个 EditText	android:id	@+id/edt_height @+id/edt_weight	设置组件 id
		android:layout_width	match_parent	宽度匹配父容器
		android:layout_height	wrap_content	高度适合内容即可
		android:inputType	numberDecimal	只能输入浮点型数据
5	Button	android:id	@+id/btn_calc	设置组件 id
		android:layout_width	wrap_content	宽度适合内容即可
		android:layout_height	wrap_content	高度适合内容即可
		android:text	计算	设置提示文本
		android:onClick	clickCalc	单击事件名称
		android:layout_gravity	center	在容器中居中对齐

参考代码如下：

```xml
<LinearLayout xmlns:android="http://schemas.android.com/apk/res/android"
    android:layout_width="match_parent"
    android:layout_height="match_parent"
    android:orientation="vertical" >
<LinearLayout
        android:layout_width="match_parent"
    android:layout_height="wrap_content"
    android:orientation="horizontal" >
     <TextView
    android:id="@+id/tv_height"
    android:layout_width="wrap_content"
    android:layout_height="wrap_content"
    android:text="身高(m)"
    android:textAppearance="?android:attr/textAppearanceLarge"/>
     <EditText
            android:id="@+id/edt_height"
            android:layout_width="match_parent"
            android:layout_height="wrap_content"
            android:ems="10"
            android:inputType="numberDecimal" >
<requestFocus />
     </EditText>
</LinearLayout>
<LinearLayout
        android:layout_width="match_parent"
    android:layout_height="wrap_content"
    android:orientation="horizontal" >
     <TextView
            android:id="@+id/tv_weight"
            android:layout_width="wrap_content"
            android:layout_height="wrap_content"
            android:text="体重(kg)"
            android:textAppearance="?android:attr/textAppearanceLarge"/>
     <EditText
            android:id="@+id/edt_weight"
            android:layout_width="match_parent"
            android:layout_height="wrap_content"
            android:inputType="numberDecimal"/>
</LinearLayout>
<Button
        android:id="@+id/btn_calc"
        android:layout_width="wrap_content"
        android:layout_height="wrap_content"
        android:text="计算"
        android:onClick="clickCalc"
        android:layout_gravity="center"/>
</LinearLayout>
```

（3）定义 result_layout.xml 布局

本布局只需要用到 1 个文本组件，属性如表 5-2 所示，代码比较简单，此处不再给出。

表 5-2 "身体质量指数测试"中各组件属性列表

序号	组件类型	属性名称	属性值	说明
1	LinearLayout	android:layout_width	match_parent	宽度匹配父容器
		android:layout_height	match_parent	高度匹配父容器
		android:orientation	vertical	垂直方向
2	TextView	android:id	@+id/tv_result	宽度匹配父容器
		android:layout_width	wrap_content	宽度适合内容即可
		android:layout_height	wrap_content	高度适合内容即可
		android:textAppearance	?android:attr/textAppearanceMedium	设置中号文本显示
		android:textColor	#000	设置文本颜色

（4）实现 MainActivity 的功能

MainActivity 类主要实现两个功能：第一，获取用户输入的数值，并计算 BMI；第二，单击"计算"按钮时，将计算出的 BMI 值传递给 ResultActivity.java 类，并显示 ResultActivity 界面。

从一个 Activity 切换到另外一个 Activity 时需要定义 Intent（意图），并通过调用 Intent 类的 putExtra(String name, float value)方法将 BMI 值传递给 ResultActivity.java。

需要注意的是，由于在计算 BMI 时用到了除法运算，为了避免出现除数为 0 的异常，需要运用到 try 语句。

参考代码如下：

```java
public class MainActivity extends Activity {
    EditText edtHeight, edtWeight;
    @Override
    protected void onCreate(Bundle savedInstanceState) {
        super.onCreate(savedInstanceState);
        setContentView(R.layout.activity_main);
        //定义组件
        edtHeight = (EditText) findViewById(R.id.edt_height);
        edtWeight = (EditText) findViewById(R.id.edt_weight);
    }
    //实现单击事件
    public void clickCalc(View v) {
        try {
            //取出文本框中的数据
            float height = (Float.valueOf(edtHeight.getText().toString()));
            float weight = (Float.valueOf(edtWeight.getText().toString()));
            //计算 BMI
            float result = weight/(height * height);
            //定义意图
            Intent i = new Intent(MainActivity.this, ResultActivity.class);
```

```
            //传递 BMI 结果数据
            i.putExtra("result",result);
            //启动意图
            startActivity(i);
        }catch(Exception e){
            //捕获异常
            Toast.makeText(MainActivity.this,"请您输入正确的数值",
                    Toast.LENGTH_SHORT).show();
        }
    }
}
```

(5) 定义 ResultActivity.java 类

ResultActivity 类中需要取得由 MainActivity 类传递的数据,并根据数值范围确定提示文本。

参考代码如下:

```
public class ResultActivity extends Activity {
    TextView tvResult;
    @Override
    protected void onCreate(Bundle savedInstanceState) {
        super.onCreate(savedInstanceState);
        this.setContentView(R.layout.result_layout);
        tvResult = (TextView) findViewById(R.id.tv_result);
        //取得传递的数据
        float result = getIntent().getExtras().getFloat("result");
        String resultText;
        //判断数据范围,确定提示文本
        if(result < 18.5){
            resultText = "您的 BMI 偏低,要注意了啊!";
        }else if(result > 24){
            resultText = "您的 BMI 太高了,需要瘦身啦!";
        }else{
            resultText = "恭喜您,您的 BMI 指数正常,继续保持噢!";
        }
        tvResult.setText(resultText);
    }
}
```

(6) 配置项目 AndroidManifest.xml

在任何一个项目中,需要显示的 Activity 类均需要在 AndroidManifest.xml 中用 <activity/> 标签进行声明配置,其中 android:name 属性即为 Activity 类的名称。

参考代码如下:

```
<?xml version="1.0" encoding="utf-8"?>
<manifest xmlns:android="http://schemas.android.com/apk/res/android"
    package="com.book.testintent"
```

```
            android:versionCode = "1"
            android:versionName = "1.0" >
    <uses-sdk
            android:minSdkVersion = "8"
            android:targetSdkVersion = "18"/>
    <application
            android:allowBackup = "true"
            android:icon = "@drawable/ic_launcher"
            android:label = "@string/app_name"
            android:theme = "@style/AppTheme" >
        <activity
                android:name = "MainActivity"
                android:label = "@string/app_name" >
    <intent-filter>
    <action android:name = "android.intent.action.MAIN"/>
    <category android:name = "android.intent.category.LAUNCHER"/>
    </intent-filter>
    </activity>
                <!-- 将结果界面使用对话框主题 -->
    <activity
            android:name = "ResultActivity"
            android:theme = "@android:style/Theme.Dialog"
            android:label = "身体质量指数"/>
    </application>
    </manifest>
```

5.1.2 活动（Activity）的定义与使用

1. Activity 的定义

在 Android 中 Activity 代表了手机屏幕的一屏，或是平板电脑中的一个窗口，被译为"活动"。它是 Android 应用最重要的组成单元之一，提供了和用户交互的可视化界面。此类定义如下：

```
java.lang.Object
  ↳ android.content.Context
  ↳ android.content.ContextWrapper
  ↳ android.view.ContextThemeWrapper
  ↳ android.app.Activity
```

创建 Activity 时，需要让类继承于 android.app.Activity 并至少实现父类的 onCreate()方法。在 onCreate()方法中需要调用 setContentView()方法设置 Activity 的使用布局。

参考代码如下：

```
public class ResultActivity extends Activity {
    @Override
    protected void onCreate(Bundle savedInstanceState) {
        super.onCreate(savedInstanceState);
```

```
                this.setContentView(R.layout.result_layout);
        }
    }
```

在创建了新的 Activity 时，一定要在 AndroidManifest.xml 中进行声明才可以使用，参考代码如下：

```
<!--将结果界面使用对话框主题-->
<activity
        android:name = "ResultActivity"
        android:theme = "@android:style/Theme.Dialog"
        android:label = "身体质量指数"/>
```

在 XML 文件中，使用 android:name 属性设置 Activity 的类名，android:theme 定义其使用的主题格式，android:label 定义标题。

2. Activity 的生命周期

在 Android 应用中，可以有多个 Activity，它们组成了 Activity 栈。当前活动的 Activity 位于栈顶，之前的 Activity 被压入下面，成为非活动状态的 Activity，他们等待是否可能被恢复为活动状态。

Activity 有 4 个重要的状态，分别是活动状态、暂停状态、停止状态和销毁状态，具体如表 5-3 所示。

表 5-3　Activity 的重要状态列表

序号	状态名称	说明
1	活动状态	当前的 Activity，位于 Activity 栈的栈顶，用户可见，并且可以获得焦点
2	暂停状态	失去焦点的 Activity，仍然可见。在 HoneyComb 版本之前，系统内存低的情况下，Activity 可能被系统杀死，所以如果有必要的话，需要重写在此方法，在其中保存用户的数据
3	停止状态	该 Activity 被其他 Activity 覆盖，不可见，但它仍然保存所有的状态和信息。当内存低时，将要被系统杀死
4	销毁状态	该 Activity 结束，或 Activity 所在的 Dalvik 进程结束

官方文档中给出了 Activity 各种状态间的关系和回调方法，如图 5-2 所示。

3. Activity 的 3 个嵌套生命周期循环

① Activity 的完整生命周期。自第一次调用 onCreate() 开始，直至调用 onDestroy() 为止。Activity 在 onCreate() 中设置所有"全局"状态以完成初始化，而在 onDestroy() 中释放所有系统资源。

② Activity 的可视生命周期。自 onStart() 调用开始直到相应的 onStop() 调用结束。在此期间，用户可以在屏幕上看到 Activity，尽管它也许并不是位于前台或者也不与用户进行交互。在这两个方法之间，我们可以保留用来向用户显示这个 Activity 所需的资源。

③ Activity 的前台生命周期。自 onResume() 调用开始到相应的 onPause() 调用结束。在此期间，Activity 位于前台最上面并与用户进行交互。Activity 会经常在暂停和恢复之间进行状态转换。例如当设备转入休眠状态或者有新的 Activity 启动时，将调用 onPause() 方法；当 Activity 获得结果或者接收到新的 Intent 时会调用 onResume() 方法。

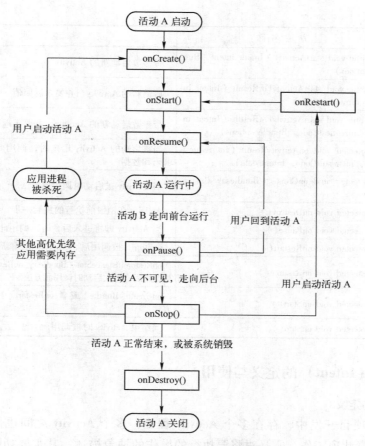

图 5-2　Activity 的生命周期图

4. Activity 的常用方法

Activity 中的常用方法如表 5-4 所示。

表 5-4　Activity 的常用方法说明

序　号	方 法 名 称	说　　明
1	public View findViewById（int id）	查找对应 id 的组件
2	public void finish( )	结束当前 Activity
3	public Intent getIntent( )	获取打开当前 Activity 的意图
4	public SharedPreferences getPreferences（int mode）	获取 SharedPreferences 对象，以便访问 Activity 可以使用的私有数据
5	public Object getSystemService（String name）	根据给定参数名称获取相应的系统服务
6	public void setContentView（int layoutResID）	用布局资源设置界面
7	public void setContentView（View view）	用定义的 View 设置界面
8	public void setIntent（Intent newIntent）	设置由 getIntent( ) 取得的意图
9	public void startActivity（Intent intent）	启动新的 Activity

(续)

序号	方法名称	说明
10	public void startActivity（Intent intent, Bundle options）	带参数启动新的 Activity
11	public void startActivityForResult（Intent intent, intrequestCode）	启动新的 Activity，并要求返回值
12	public void startActivityForResult（Intent intent, intrequestCode, Bundle options）	带参数启动新的 Activity，并要求返回值
13	protected void onActivityResult（int requestCode, int resultCode, Intent data）	当被启动的 Activity 退出时，调用此方法，同时带回请求码和数据
14	protected void onCreate（BundlesavedInstanceState）	当 Activity 被启动时调用此方法
15	protected void onDestroy（）	Activity 在销毁前执行的最后一步
16	protected void onPause（）	当 Activity 即将进入后台时，调用此方法
17	protected void onRestart（）	onStop（）被调用后，当 Activity 重新显示时调用此方法
18	protected void onResume（）	onRestoreInstanceState（Bundle），onRestart（）或者 onPause（）被调用后，再次启动时调用此方法
19	protected void onStart（）	onCreate（Bundle）或者 onRestart（）被调用后，调用此方法
20	protected void onStop（）	当结束 Activity 时被调用此方法

5.1.3 意图（Intent）的定义与使用

1. Intent 的定义

在 Android 项目开发中，存在多个 Activity 程序，多个 Activity 之间也需要进行相互通信。Intent 就负责此项工作，它是对将要执行的操作的抽象描述。其主要功能在于完成多个 Activity 之间进行切换，切换时还可以携带必要的数据。此类定义如下：

```
java.lang.Object
  ↳ android.content.Intent
```

可以使用该类的构造函数创建 Intent，参考代码如下：

```
Intent i = new Intent( MainActivity.this, ResultActivity.class);
```

上述代码中使用到了 Intent 的 Intent（Context packageContext，Class<?> cls）构造函数，其中有两个参数，第 1 个表示当前环境为 MainActivity.this，第 2 个表示目标类为 ResultActivity.class。

2. 使用 Intent 传递数据

创建好意图后，还可通过调用 Intent 类的 putExtra（）方法添加携带数据。参考代码如下：

```
i.putExtra("result", result);
```

在将意图动作和数据都准备好之后，即可调用 Activity 的 startActivity（）方法启动意图，实现通信。参考代码如下：

```
startActivity(i);
```

3. Intent 的常用方法

Activity 中的常用方法如表 5-5 所示。

表 5-5　Activity 的常用方法说明

序号	方法名称	说明
1	public Intent()	创建空的 Intent
2	public Intent(Intenti)	复制 Intent 对象
3	public Intent(String action)	指定跳转的 Activity 名称
4	public Intent(String action, Uri uri)	指定跳转的 Activity 名称和传递的 URI 信息
5	public Intent(Context packageContext, Class<?> cls)	指定操作的上下文以及跳转的 Activity
6	public Intent addCategory(String category)	增加 Category 数据
7	public Intent addFlags(int flags)	增加 flag 标记
8	public boolean[] getBooleanArrayExtra(String name)	获取布尔数组的值
9	public boolean getBooleanExtra(String name, boolean defaultValue)	获取布尔变量的值
10	public Bundle getBundleExtra(String name)	获取 Bundle 数据
11	public byte[] getByteArrayExtra(String name)	获取字节数组的值
12	public byte getByteExtra(String name, byte defaultValue)	获取字节变量的值
13	public Set<String> getCategories()	获取类别信息
14	public ComponentName getComponent()	获取组件信息
15	public Uri getData()	获取 Uri 信息
16	public double[] getDoubleArrayExtra(String name)	获取 double 数组的值
17	public double getDoubleExtra(String name, double defaultValue)	获取 double 变量的值
18	public Bundle getExtras()	获取所有附加信息
19	public int getFlags()	获取设置的标记信息
20	public int[] getIntArrayExtra(String name)	获取整型数组的值
21	public int getIntExtra(String name, int defaultValue)	获取整型变量的值
22	public String getStringExtra(String name)	获取字符串的值
23	public String getType()	获取 MIME 类型
24	public boolean hasCategory(String category)	判断是否有指定的 Category
25	public boolean hasExtra(String name)	判断是否有指定的附加信息
26	public Intent putExtra(String name, double[] value)	为附加信息添加 double 数组
27	public Intent putExtra(String name, int value)	为附加信息添加 int 数据
28	public Intent putExtra(String name, Bundle value)	为附加信息添加 Bundle 数据
29	public Intent putExtra(String name, Serializable value)	设置一个可序列化对象的附加信息
30	public Intent putExtra(String name, int[] value)	为附加信息添加 int 数组
31	public Intent putExtra(String name, float value)	为附加信息添加 float 数据
32	public Intent putExtra(String name, byte[] value)	为附加信息添加字节数组
33	public Intent putExtra(String name, boolean value)	为附加信息添加 boolean 数组
34	public Intent putExtra(String name, String value)	为附加信息添加 String 数组
35	public Intent putExtras(Bundle extras)	为附加信息添加 Bundle 信息

(续)

序号	方法名称	说明
36	public void removeCategory(String category)	删除指定的 Category 数据信息
37	public void removeExtra(String name)	删除指定的附加信息
38	public Intent setAction(String action)	设置操作名称
39	public Intent setClass(Context packageContext, Class<?> cls)	设置要跳转的 Activity
40	public Intent setComponent(ComponentName component)	设置目标组件
41	public Intent setData(Uri data)	设置操作的 URI
42	public Intent setDataAndType(Uri data, String type)	设置数据并指定 MIME 类型
43	public Intent setFlags(int flags)	设置标记
44	public Intent setType(String type)	设置数据的 MIME 类型

4. Intent 可以传递的数据种类

使用 Intent 传递的数据可以分为 7 种类型，分别是操作（Action）、数据（Data）、数据类型（Type）、操作类别（Category）、附加信息（Extras）、组件（Component）和标志（Flags）。

通过调用 setAction() 方法可以设置 Intent 会触发的操作类型。系统常用的 Action 如表 5-6 所示。

表 5-6 系统常用的 Action 列表

序号	名称	AndroidManifest.xml 中的配置名称	说明
1	ACTION_MAIN	android.intent.action.MAIN	作为一个程序的入口，不需要接收数据
2	ACTION_VIEW	android.intent.action.VIEW	用于数据的显示
3	ACTION_DIAL	android.intent.action.DIAL	调用拨号程序
4	ACTION_EDIT	android.intent.action.EDIT	用于编辑给定的数据
5	ACTION_PICK	android.intent.action.PICK	从特定的数据中选择
6	ACTION_RUN	android.intent.action.RUN	运行数据
7	ACTION_SENDTO	android.intent.action.SENDTO	调用发送短信程序
8	ACTION_GET_CONTENT	android.intent.action.GET_CONTENT	根据指定的 Type 来选择打开操作内容的 Intent
9	ACTION_CHOOSER	android.intent.action.CHOOSER	创建文件操作选择器

通过调用 setData() 方法可以设置 Intent 所操作数据的 URI 类型。不同的操作对应不同的 Data，常用的数据格式如表 5-7 所示。

表 5-7 常用的数据格式列表

序号	Data（URI）格式	说明
1	http:///网页格式	浏览网页
2	tel：电话号码	拨打电话
3	smsto：短信接收人号码	发送短信
4	file:///sdcard/文件或目录	查找 SD 卡文件

通过调用 addCategory( )方法可以设置多个类别。常用的 Category 如表 5-8 所示。

表 5-8 常用的 Category 列表

序号	Category 名称	AndroidManifest.xml 中的配置名称	说明
1	CATEGORY_LAUNCHER	android.intent.category.LAUNCHER	此程序显示在应用程序列表中
2	CATEGORY_HOME	android.intent.category.HOME	开机时的第 1 个界面
3	CATEGORY_DEFAULT	android.intent.category.DEFAULT	设置操作默认执行
4	CATEGORY_PREFERENCE	android.intent.category.PREFERENCE	运行后出现一个选择面板

附加信息可以使用 putExtra( )方法进行设置,主要功能是传递应用程序所需要的一些额外信息,通常是一组键值对。

组件指明了将要处理的 Activity 程序,所有的组件信息都被封装在一个 ComponentName 对象中,这些组件都必须在 AndroidManifest.xml 文件的 <application> 中注册。

标志用于指示 Android 系统如何加载并运行一个操作,可以通过 addFlags( )方法添加。

数据类型指定要传送数据的 MIME 类型,可以直接通过 setType( )方法进行设置。

5.2 实例 2:编辑商品信息

5.2.1 功能要求与操作步骤

1. 功能要求

完成如图 5-3 所示的应用程序。要求能够:

① 图 5-3a 所示的界面可以显示有关商品的信息,单击下方的"编辑商品信息"按钮时,会弹出如图 5-3b 所示的界面,其中各个可编辑组件的初始值均由 5-3a 传递。

② 图 5-3b 中所有文本框均可被编辑,单击图片框后可以更改图片,如图 5-3c 所示。

③ 在用户将商品的所有信息编辑完毕,单击 5-3c 下方的"保存查看"按钮后,可以将更新后的信息显示在首界面,如图 5-3d 所示。

2. 操作步骤

(1) 创建项目

参考如下输入信息创建 Android 项目:

```
Application Name:05_TestGetData
Project Name:05_TestGetData
Package Name:com.book.testgetdata
Activity Name:MainActivity
Layout Name:activity_main
```

(2) 定义显示信息布局 activity_main.xml

从图 5-3a 可看出,应用布局使用到了 10 个 TextView、1 个 ImageView 和 1 个 Button。布局的根元素用垂直线性布局,其中第 1 个元素即为显示文本"商品信息",第 2 个元素是嵌套的表格布局,用来显示商品的详细信息,第 3 个元素是"编辑商品信息"按钮。各个

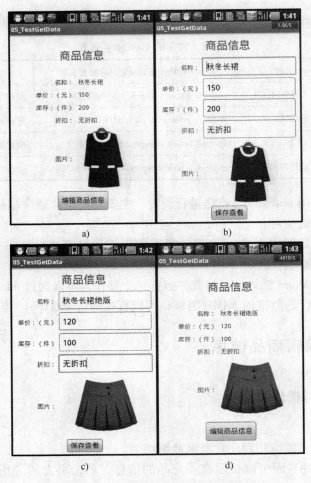

图5-3 "编辑商品信息"运行效果图

组件的属性可参考表5-9设置。

表5-9 "编辑商品信息"中各组件属性列表

序号	组件类型	属性名称	属性值	说明
1	LinearLayout	android:layout_width	match_parent	宽度匹配父容器
		android:layout_height	match_parent	高度匹配父容器
		android:orientation	vertical	垂直方向
2	最上方的TextView	android:layout_width	match_parent	宽度匹配父容器
		android:layout_height	wrap_content	高度适合内容
		android:text	商品信息	显示文本
		android:textSize	25sp	设置字号
		android:gravity	center	设置对齐方式
		android:padding	30dp	设置填充值

（续）

序号	组件类型	属性名称	属性值	说明
3	TableLayout	android:layout_width	match_parent	宽度匹配父容器
		android:layout_height	wrap_content	高度适合内容
4	5个 TableRow	android:layout_width	wrap_content	宽度适合内容
		android:layout_height	wrap_content	高度适合内容
		android:gravity	center	设置对齐方式
5	第1列的5个 TextView	android:layout_width	wrap_content	宽度适合内容
		android:layout_height	wrap_content	高度适合内容
		android:text	图片： 折扣： 库存： 单价： 名称：	设置显示文本
		android:gravity	right	设置对齐方式
		android:padding	5dp	设置填充值
6	第2列的4个 TextView	android:id	@+id/tv_discount @+id/tv_store @+id/tv_price @+id/tv_name	设置组件 id
		android:layout_width	wrap_content	宽度适合内容
		android:layout_height	wrap_content	高度适合内容
		android:text	无折扣 200 150 秋冬长裙	设置显示文本
		android:padding	5dp	设置填充值
7	ImageView	android:id	@+id/img_pic	设置组件 id
		android:layout_width	wrap_content	宽度适合内容
		android:layout_height	wrap_content	高度适合内容
		android:src	@drawable/skirt1	设置显示图片
8	Button	android:layout_width	wrap_content	宽度适合内容
		android:layout_height	wrap_content	高度适合内容
		android:text	编辑商品信息	设置显示文本
		android:layout_gravity	center	设置对齐方式
		android:onClick	clickEdit	设置单击事件名称

参考代码如下：

```
< LinearLayout xmlns:android = " http://schemas.android.com/apk/res/android"
    android:layout_width = " fill_parent"
    android:layout_height = " fill_parent"
    android:orientation = " vertical" >
```

```xml
<TextView
        android:layout_width = "match_parent"
        android:layout_height = "wrap_content"
        android:text = "商品信息"
        android:textSize = "25sp"
        android:gravity = "center"
        android:padding = "30dp"/>
<TableLayout
        android:layout_width = "match_parent"
        android:layout_height = "wrap_content" >
    <TableRow
            android:layout_width = "match_parent"
            android:layout_height = "wrap_content"
            android:gravity = "center" >
        <TextView
                android:layout_width = "match_parent"
                android:layout_height = "wrap_content"
                android:text = "名称:"
                android:gravity = "right"
                android:padding = "5dp"/>
        <TextView
                android:id = "@+id/tv_name"
                android:layout_width = "match_parent"
                android:layout_height = "wrap_content"
                android:text = "秋冬长裙"
                android:padding = "5dp"/>
    </TableRow>
    <TableRow
            android:layout_width = "match_parent"
            android:layout_height = "wrap_content"
            android:gravity = "center" >
        <TextView
                android:layout_width = "wrap_content"
                android:layout_height = "wrap_content"
                android:text = "单价:(元)"
                android:padding = "5dp"/>
        <TextView
                android:id = "@+id/tv_price"
                android:layout_width = "wrap_content"
                android:layout_height = "wrap_content"
                android:text = "150"
                android:padding = "5dp"/>
    </TableRow>
    <TableRow
            android:layout_width = "wrap_content"
            android:layout_height = "wrap_content"
            android:gravity = "center" >
        <TextView
```

```xml
            android:layout_width = "wrap_content"
            android:layout_height = "wrap_content"
            android:text = "库存:(件)"
            android:padding = "5dp"/>
<TextView
            android:id = "@+id/tv_store"
            android:layout_width = "wrap_content"
            android:layout_height = "wrap_content"
            android:text = "200"
            android:padding = "5dp"/>
</TableRow>
<TableRow
            android:layout_width = "wrap_content"
            android:layout_height = "wrap_content"
            android:gravity = "center" >
<TextView
            android:layout_width = "wrap_content"
            android:layout_height = "wrap_content"
            android:text = "折扣:"
            android:gravity = "right"
            android:padding = "5dp"/>
<TextView
            android:id = "@+id/tv_discount"
            android:layout_width = "wrap_content"
            android:layout_height = "wrap_content"
            android:text = "无折扣"
            android:padding = "5dp"/>
</TableRow>
<TableRow
            android:layout_width = "wrap_content"
            android:layout_height = "wrap_content"
            android:gravity = "center" >
<TextView
            android:layout_width = "wrap_content"
            android:layout_height = "wrap_content"
            android:text = "图片:"
            android:gravity = "right"
            android:padding = "5dp"/>
<ImageView
            android:id = "@+id/img_pic"
            android:layout_width = "wrap_content"
            android:layout_height = "wrap_content"
            android:src = "@drawable/skirt1"/>
</TableRow>
</TableLayout>
<Button
        android:layout_width = "wrap_content"
        android:layout_height = "wrap_content"
```

```
            android:text = "编辑商品信息"
            android:layout_gravity = "center"
            android:onClick = "clickEdit" />
    </LinearLayout >
```

（3）定义修改信息布局 edit_layout. xml

从图 5-3b 可看出，edit_layout. xml 与 activity_main. xml 的布局很相似，只是将表格布局中第 2 列的"TextView"替换为"EditText"；将下方按钮的显示文本设置为"保存查看"即可。为了使图片框能响应单击事件，需要为图片框添加 android:onClick = "clickSave" 属性。

（4）在 MainActivity 中实现显示信息等功能

MainActivity 的主要功能包含：第一，显示商品原有信息；第二，单击"编辑商品信息"按钮时将商品的原有信息使用 startActivityForResult(Intent intent, int requestCode) 方法传递到修改界面；第三，使用 onActivityResult(int requestCode, int resultCode, Intent i) 方法接收修改界面的新数据并更新商品信息。

参考代码如下：

```
public class MainActivity extends Activity {
    //声明组件
    TextView tvName, tvPrice, tvStore, tvDis;
    ImageView imgPic;
    //声明图片资源变量
    int imgRes;
    @Override
    protected void onCreate( Bundle savedInstanceState) {
        super.onCreate( savedInstanceState) ;
        setContentView( R. layout. activity_main) ;
        //找组件
        tvName = ( TextView) this. findViewById( R. id. tv_name) ;
        tvPrice = ( TextView) this. findViewById( R. id. tv_price) ;
        tvStore = ( TextView) this. findViewById( R. id. tv_store) ;
        tvDis = ( TextView) this. findViewById( R. id. tv_discount) ;
        imgPic = ( ImageView) this. findViewById( R. id. img_pic) ;
    }
    public void clickEdit( View v) {
        //获取图片资源 ID
        if( imgPic. getDrawable( ). getConstantState( ). equals(
            getResources( ). getDrawable( R. drawable. skirt1). getConstantState( ) ) ) {
            imgRes = R. drawable. skirt1;
        } else {
            imgRes = R. drawable. skirt2;
        }
        //定义意图
        Intent i = new Intent( MainActivity. this, EditActivity. class) ;
        //携带数据
        i. putExtra( "name", tvName. getText( ). toString( ) ) ;
        i. putExtra( "price", tvPrice. getText( ). toString( ) ) ;
```

```java
                i.putExtra("store",tvStore.getText().toString());
                i.putExtra("discount",tvDis.getText().toString());
                i.putExtra("img",imgRes);
                //启动意图,并要求返回数据,requestCode 为 0
                startActivityForResult(i,0);
            }
            @Override
            protected void onActivityResult(int requestCode,int resultCode,Intent i){
                //在有数据正确返回时,将所有信息更新为新数据
                if(resultCode==RESULT_OK){
                    tvName.setText(i.getStringExtra("name"));
                    tvPrice.setText(i.getStringExtra("price"));
                    tvStore.setText(i.getStringExtra("store"));
                    tvDis.setText(i.getStringExtra("discount"));
                    imgPic.setImageResource(i.getIntExtra("img",R.drawable.skirt1));
                }
            }
        }
```

（5）在 EditActivity 中实现修改信息等功能

EditActivity 的主要功能包含：第一，将商品原有信息显示在对应的文本框中；第二，单击图片框时更改显示图片；第三，单击"保存查看"按钮时更新商品信息并将新数据带回 MainActivity。

参考代码如下：

```java
        public class EditActivity extends Activity {
            //声明组件
            EditText edtName,edtPrice,edtStore,edtDiscount;
            ImageView imgEdtPic;
            //定义意图
            Intent i = null;
            //声明图片资源 ID
            int imgRes;
            @Override
            protected void onCreate(Bundle savedInstanceState) {
                super.onCreate(savedInstanceState);
                this.setContentView(R.layout.edit_layout);
                //找组件
                edtName = (EditText) this.findViewById(R.id.edt_name);
                edtPrice = (EditText) this.findViewById(R.id.edt_price);
                edtStore = (EditText) this.findViewById(R.id.edt_store);
                edtDiscount = (EditText) this.findViewById(R.id.edt_discount);
                imgEdtPic = (ImageView) this.findViewById(R.id.img_edt_pic);
                //取得源意图
                i = getIntent();
                //获取源意图中的数据
                edtName.setText(i.getStringExtra("name"));
```

```java
            edtPrice.setText(i.getStringExtra("price"));
            edtStore.setText(i.getStringExtra("store"));
            edtDiscount.setText(i.getStringExtra("discount"));
            imgRes = i.getIntExtra("img",R.drawable.skirt1);
            imgEdtPic.setImageResource(imgRes);
    }
    //单击图片框的事件
    public void clickImg(View v){
        //更改 imgRes 值
        if(imgEdtPic.getDrawable().getConstantState().equals(
            getResources().getDrawable(R.drawable.skirt1).getConstantState())){
            imgRes = R.drawable.skirt2;
        }else{
            imgRes = R.drawable.skirt1;
        }
        //更改显示的图片
        imgEdtPic.setImageResource(imgRes);
    }
    //单击"保存查看"按钮事件
    public void clickSave(View v){
        //携带新值
        i.putExtra("name",edtName.getText().toString());
        i.putExtra("price",edtPrice.getText().toString());
        i.putExtra("store",edtStore.getText().toString());
        i.putExtra("discount",edtDiscount.getText().toString());
        i.putExtra("img",imgRes);
        //返回结果
        this.setResult(RESULT_OK,i);
        finish();
    }
}
```

5.2.2 获取 Activity 返回值的方法

1. 源 Activity 调用 startActivityForResult（Intent intent，int requestCode）方法启动意图

源 Activity 需要调用 startActivityForResult(Intent intent,int requestCode)方法启动意图,才能获取到目标 Activity 中返回的数据,参数分别是意图和请求码,其中请求码可以自定义。参考代码如下：

```java
Intent i = new Intent(MainActivity.this,EditActivity.class);
startActivityForResult(i,0);
```

2. 目标 Activity 获取意图源更新数据并结束 Activity 运行

如果目标 Activity 需要将数据返回源 Activity,那么首先需要调用 getIntent()取得源意图,然后将需要返回的数据携带,并调用 setResult(int resultCode,Intent data)和 finish()方法设置返回结果, setResult 方法的参数为结果码（RESULT_OK）和源意图。参考代码如下：

```
        i = getIntent();
        i.putExtra("name",edtName.getText().toString());
        i.putExtra("price",edtPrice.getText().toString());
        i.putExtra("store",edtStore.getText().toString());
        setResult(RESULT_OK,i);
        finish();
```

3. 源 Activity 重写 onActivityResult(int requestCode,int resultCode,Intent i)方法

源 Activity 如果需要接收返回的数据，需要重写父类的 onActivityResult(int requestCode, int resultCode,Intent i)方法，参数分别是请求码，结果码和意图。参考代码如下：

```
    @Override
        protected void onActivityResult(int requestCode,int resultCode,Intent i) {
            //在有数据正确返回时，将所有信息更新为新数据
            if(resultCode == RESULT_OK) {
                tvName.setText(i.getStringExtra("name"));
                tvPrice.setText(i.getStringExtra("price"));
            }
        }
```

5.3 实例3：快速联系

5.3.1 功能要求与操作步骤

1. 功能要求

完成如图 5-4 所示的应用程序。要求能够：

① 应用程序的主界面如图 5-4a 所示。用户输入联系人姓名的第 1 个字符后可以显示相关列表，效果如图 5-4b 所示。

② 用户选择了列表项中的联系人之后，单击"拨打电话"按钮即可拨号，效果如图 5-4c 所示；单击"发送短信"按钮即可发短信，效果如图 5-4d 所示。

2. 操作步骤

（1）创建项目

参考如下输入信息创建 Android 项目：

```
Application Name:05_TestSystemIntent
Project Name:05_TestSystemIntent
Package Name:com.book.testsystemintent
Activity Name:MainActivity
Layout Name:activity_main
```

（2）定义布局

"快速联系"应用程序只有一个界面布局 activity_main.xml。在这个布局中需要用到 1 个 TextView 显示文本框、1 个 AutoCompleteTextView 自动完成文本框和 2 个 Button 按钮。这些组件在线性布局中垂直排列即可。

图 5-4 "快速联系"运行效果图

(3) 在 MainActivity 中实现功能

MainActivity 中实现的主要功能包括：第一，联系人姓名可以自动完成；第二，单击自动完成文本框的列表项时，可以获取联系人的电话号码；第三，单击"拨打电话"按钮可以调用系统的拨号程序，实现拨号功能；第四，单击"发送短信"按钮可以调用系统的发送短信 Intent，实现短信发送。

参考代码如下：

```java
public class MainActivity extends Activity implements OnItemClickListener {
    //联系人姓名列表
    List < String >  name = new ArrayList < String > ( );
    //联系人 ID 列表
    List < Integer >  contact_id = new ArrayList < Integer > ( );
    //声明联系人的电话
    String phone = " " ;
    //声明自动完成文本框组件
    AutoCompleteTextView actv;
    @Override
```

```java
protected void onCreate(Bundle savedInstanceState) {
    super.onCreate(savedInstanceState);
    setContentView(R.layout.activity_main);
    actv = (AutoCompleteTextView) findViewById(R.id.actv_contact);
    //获得 ContentResolver 对象
    ContentResolver cr = this.getContentResolver();
    //查询记录
    Cursor cursor = cr.query(
        Contacts.CONTENT_URI,
        null, null, null, Contacts.DISPLAY_NAME + " ASC");
    if(cursor! = null) {
        for(cursor.moveToFirst();! cursor.isAfterLast();cursor.moveToNext()) {
            //将所有联系人的姓名和 ID 存放在对应的列表中
            name.add(cursor.getString(
                    cursor.getColumnIndex(Contacts.DISPLAY_NAME)));
            contact_id.add(cursor.getInt(
                    cursor.getColumnIndex(Contacts._ID)));
        }
    }
    //用联系人姓名生成适配器
    ArrayAdapter<String> aa = new ArrayAdapter<String>(
            MainActivity.this,
            android.R.layout.simple_list_item_1,
            name);
    //为自动完成文本框设置适配器
    actv.setAdapter(aa);
    //为自动完成文本框添加列表项单击事件
    actv.setOnItemClickListener(this);
}
@Override
public void onItemClick(AdapterView<?> adapterView, View v, int pos, long id) {
    //查找第一次出现选择联系人姓名的位置
    int index = name.indexOf(adapterView.getItemAtPosition(pos));
    //设置查找电话号码的条件
    String[] selArgs = new String[] {"" + contact_id.get(index) + ""};
    //查询记录
    Cursor c = getContentResolver().query(
            Phone.CONTENT_URI,
            null,
            Phone.CONTACT_ID + " = ?",
            selArgs,
            null);
    if(! c.equals(null)) {
        c.moveToFirst();
        //取出联系人的电话号码字段值
        phone = c.getString(c.getColumnIndex(Phone.NUMBER));
    }
```

```java
        }
        //单击"拨打电话"的处理
        public void clickDial(View v){
            Intent i = new Intent();
            //指定 Action
            i.setAction(Intent.ACTION_CALL);
            //设置数据
            i.setData(Uri.parse("tel:" + phone));
            //启动 Activity
            startActivity(i);
        }
        //单击"发送短信"的处理
        public void clickSMS(View v){
            Intent i = new Intent();
            //指定 Action
            i.setAction(Intent.ACTION_SENDTO);
            //设置数据
            i.setData(Uri.parse("smsto:" + phone));
            startActivity(i);
        }
    }
```

(4) 在 AndroidManifest.xml 中设置权限

程序中用到了拨号和查看联系人的操作，需要增加拨打电话和查看联系人的权限，参考代码如下：

```xml
<uses-permission android:name="android.permission.READ_CONTACTS"/>
<uses-permission android:name="android.permission.CALL_PHONE"/>
```

5.3.2 调用拨号程序和短信程序的方法

1. 启动拨号程序的方法

拨打电话在 Android 系统中可以通过程序的调用完成。如果设置 Action 为 "ACTION_DIAL" 类型即可调用拨号程序，此时用户可以不指定拨打的电话号码，手工拨打电话；如果设置 Action 为 "ACTION_CALL" 类型即可直接拨出电话，此时需要调用 Intent 的 setData()方法结合 Uri.parse("tel:" + phone)指定拨打的电话号码。

直接拨打电话的参考代码如下：

```java
Intent i = new Intent();
i.setAction(Intent.ACTION_CALL);
i.setData(Uri.parse("tel:" + phone));
startActivity(i);
```

为了正常使用拨号程序，还需要为应用程序增加以下权限：

```xml
<uses-permission android:name="android.permission.CALL_PHONE"/>
```

2. 启动发送短信程序的方法

调用发送短信程序时,需要设置 Action 为"Intent. ACTION_SENDTO"。如果需要直接发送短信,需要调用 Intent 的 setData()方法结合 Uri. parse("smsto:" + phone)指定短信接收人的电话号码;如果不指定接收短信的电话号码则需要用户自己填写。开发人员还可以直接通过调用 Intent 的 putExtra()方法附加信息设置短信的内容,附加信息的名称为系统定义好的"sms_body"。此外,如果要发送短信,应该调用 Intent 的 setType 方法设置短信的 MIME 类型为"vnd. android – dir/mms – sms"。

参考代码如下:

```
Intent i = new Intent();
i.setAction(Intent. ACTION_SENDTO);
i.setData(Uri. parse("smsto:" + phone));
startActivity(i);
```

5.3.3 ContentProvider 共享数据的方法

1. ContentProvider 与 ContentResolver

Content Provider 用于保存和获取数据,并使其对所有应用程序可见。因为在 Android 中没有提供所有应用共同访问的公共存储区域,所以这是不同应用程序间共享数据的唯一方式。Content Provider 内部如何保存数据由其设计者决定,但是所有的 Content Provider 都实现一组通用的方法,以便用来实现数据的增、删、改、查功能。

通常客户端不会直接使用这些方法,而是通过 ContentResolver 对象实现对 Content Provider 的操作。开发人员可以通过调用 Activity 或其他应用程序组件的 getContentResolver()方法来获取 ContentResolver 对象。

参考代码如下:

```
ContentResolver cr = this. getContentResolver();
```

使用 ContentResolver 类提供的以获得 Content Provider 中的全部所需的数据。类的定义如下:

```
java. lang. Object
  ↳ android. content. ContentResolver
```

ContentResolver 类常用的方法如表 5–10 所示。

表 5–10 ContentResolver 类常用的方法说明

序 号	方 法 名 称	说 明
1	public final int delete(Uri url, String where, String[] selectionArgs)	删除数据
2	public final Uri insert(Uri url, ContentValues values)	插入数据
3	public final Cursor query(Uri uri, String[] projection, String selection, String[] selectionArgs, String sortOrder)	查询数据,返回 Cursor 对象,可以遍历各行各列来读取各个字段的值
4	public final int update(Uri uri, ContentValues values, String where, String[] selectionArgs)	更新数据

2. URI 简介

在 ContentResolver 类中常用的方法中都有 Uri 类型的参数。这是因为每个 Content Provider 都用 Uri 来唯一标识其数据集。这意味着如果开发人员自定义 Content Provider，就需要为其 URI 定义一个常量，来简化客户端代码并让日后更新更加简洁。例如，系统中存储联系人表的 URI 和存储联系人联系方式表的 URI 分别为：

```
android. provider. ContactsContract. Contacts. CONTENT_URI
android. provider. ContactsContract. CommonDataKinds. Phone. CONTENT_URI
```

3. 获取系统内置联系人和联系电话的方法

在取得 ContentResolver 对象后，即可调用 ContentResolver 类的 query(Uri uri, String[ ] projection, String selection, String[ ] selectionArgs, String sortOrder) 方法从指定的 URI 中获取需要的数据。

例如，可以获取联系人信息并按联系人姓名拼音升序排列。参考代码如下：

```
ContentResolver cr = this. getContentResolver( );
Cursor cursor = cr. query(
    Contacts. CONTENT_URI,
    null,
    null,
    null,
    Contacts. DISPLAY_NAME + " ASC" );
```

再如，可以根据联系人的 ID 取得其联系电话。参考代码如下：

```
String [ ]selArgs = new String [ ] { " " + contact_id + " " };
Cursor c = getContentResolver( ). query(
Phone. CONTENT_URI,
null,
Phone. CONTACT_ID + " = ?",
selArgs,
null);
```

查询返回的 Cursor 与类似在标准 SQL 查询操作中返回的结果集。通过它可以遍历各行各列，取得各个需要数据值。具体的操作方法可以参考操作步骤中给出的代码。

5.4 实例 4：闹钟服务

5.4.1 功能要求与操作步骤

1. 功能要求

完成如图 5-5 所示的应用程序。要求能够：

① 图 5-5a 是应用程序的主界面。用户可以在此选择闹钟时间，设置或删除闹钟。

② 当用户单击"设置闹钟"按钮后，会有如图 5-5a 所示的提示信息，并且尚未到达设置时间前，用户可以单击"删除闹钟"按钮删除此闹钟，如图 5-5b 所示。此后再单击"删

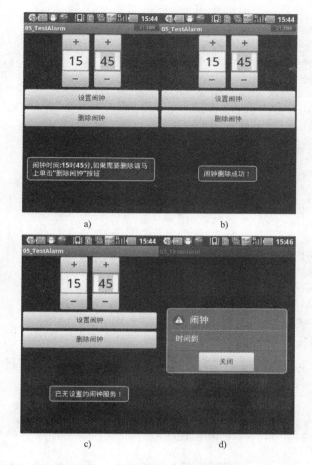

图 5-5 "闹钟服务"运行效果图

除闹钟"按钮时,有如图 5-5c 所示的提示信息。

③ 在系统时间与闹钟时间一致时,弹出如图 5-5d 的提示对话框。

2. 操作步骤

(1) 创建项目

参考如下输入信息创建 Android 项目:

```
Application Name: 05_TestAlarm
Project Name: 05_TestAlarm
Package Name: com.book.testalarm
Activity Name: MainActivity
Layout Name: activity_main
```

(2) 定义布局

"闹钟服务"应用程序的主界面中包含时间选择器和两个按钮,使用垂直线性布局即可完成布局定义。

参考代码如下:

```
<?xml version = "1.0" encoding = "utf-8"?>
<LinearLayout
```

173

```xml
xmlns:android = "http://schemas.android.com/apk/res/android"
android:orientation = "vertical"
android:layout_width = "match_parent"
android:layout_height = "match_parent" >
    <TimePicker
        android:id = "@+id/tp_time"
        android:layout_width = "match_parent"
        android:layout_height = "wrap_content"/>
    <Button
        android:layout_width = "match_parent"
        android:layout_height = "wrap_content"
        android:text = "设置闹钟"
        android:onClick = "clickSet"/>
    <Button
        android:layout_width = "match_parent"
        android:layout_height = "wrap_content"
        android:text = "删除闹钟"
        android:onClick = "clickDel"/>
</LinearLayout>
```

（3）实现 MainActivity 中的功能

MainActivity 中的主要功能包含：第一，单击"设置闹钟"按钮时，启动闹钟服务，向系统发出广播；第二，单击"删除闹钟"按钮时，取消闹钟服务。

参考代码如下：

```java
public class MainActivity extends Activity implements OnTimeChangedListener {
    // 自定义 ACTION
    public static final String ACTION_ALARM = "com.book.action.set_alarm";
    // 声明闹钟服务
    private AlarmManager am = null;
    // 声明组件
    private TimePicker tpTime;
    // 定义相关变量
    private int hourOfDay = 0;
    private int minute = 0;
    String text;
    // 获取 Calendar 对象
    private Calendar calendar = Calendar.getInstance();
    // 声明 PendingIntent 对象
    PendingIntent sender;
    @Override
    public void onCreate(Bundle savedInstanceState) {
        super.onCreate(savedInstanceState);
        super.setContentView(R.layout.activity_main);
        //获取组件
        tpTime = (TimePicker) super.findViewById(R.id.tp_time);
        //设置时间选择器为 24 小时
```

```java
        tpTime.setIs24HourView(true);
        //为时间选择器添加时间改变事件
        tpTime.setOnTimeChangedListener(this);
        //设置广播接收器
        Intent i = new Intent(MainActivity.this,AlarmReceiver.class);
        //为广播接收器设置 Action
        i.setAction(ACTION_ALARM);
        //定义附加 Intent
        sender = PendingIntent.getBroadcast(
                MainActivity.this,0,i,PendingIntent.FLAG_UPDATE_CURRENT);
        //取得当前时间
        hourOfDay = tpTime.getCurrentHour();
        minute = tpTime.getCurrentMinute();
}
//单击"设置"按钮的事件
public void clickSet(View v){
        //获取闹钟管理器
        am = (AlarmManager) getSystemService(Context.ALARM_SERVICE);
        //启动闹钟服务
        am.set(AlarmManager.RTC_WAKEUP,calendar.getTimeInMillis(),sender);
        //格式化输出信息使用的参数
        Object[] args = {hourOfDay,minute};
        //如果时、分不足两位数,用0补齐
        if(hourOfDay < 10){
            args[0] = "0" + hourOfDay;
        }
        if(minute < 10){
            args[1] = "0" + minute;
        }
        //{0}表示 args[0],{1}表示 args[1]
        text = MessageFormat.format(
            "闹钟时间:{0}时{1}分,如果需要删除请马上单击"删除闹钟"按钮",args);
        Toast.makeText(MainActivity.this,text,Toast.LENGTH_LONG).show();
}
public void clickDel(View v){
        if (am != null){
            // 取消闹钟
            am.cancel(sender);
            am = null;
            text = "闹钟删除成功!";
            Toast.makeText(MainActivity.this,text,Toast.LENGTH_LONG).show();
        }else{
            text = "已无设置的闹钟服务!";
            Toast.makeText(MainActivity.this,text,Toast.LENGTH_LONG).show();
        }
}
@Override
```

```java
        public void onTimeChanged(TimePicker view,int h,int m) {
            //当时间改变时设置 calendar 的值
            calendar.setTimeInMillis(System.currentTimeMillis());
            calendar.set(Calendar.HOUR_OF_DAY,h);
            calendar.set(Calendar.MINUTE,m);
            calendar.set(Calendar.SECOND,0);
            calendar.set(Calendar.MILLISECOND,0);
            hourOfDay = h;
            minute = m;
        }
    }
```

(4) 实现广播接收器类 AlarmReceiver.java

广播程序运行在后台,当用户设置闹钟后,闹钟的处理应用会直接将它提交给广播处理类完成,用户可以在广播接收类中进行更多的操作。通过广播跳转到 AlarmMessage 程序时,需要传递一个重要的标记:FLAG_ACTIVITY_NEW_TASK,如果没有此标记,即使广播时间到了,也不会执行指定的 Activity 程序。

参考代码如下:

```java
public class AlarmReceiver extends BroadcastReceiver {
    @Override
    public void onReceive(Context context,Intent intent) {
        //当收到广播时,弹出 AlarmMessage 对话框
        Intent it = new Intent(context,AlarmMessage.class);
        it.addFlags(Intent.FLAG_ACTIVITY_NEW_TASK);
        context.startActivity(it);
    }
}
```

(5) 实现闹钟的提示类 AlarmMessage.java

本 Activity 程序作为闹钟服务到时的信息显示类,主要功能是显示一个对话框,并且提示用户"设置的闹钟已到",而当用户单击"关闭"按钮之后,本 Activity 将直接利用 finish()方法结束。

参考代码如下:

```java
public class AlarmMessage extends Activity {
    @Override
    protected void onCreate(Bundle savedInstanceState) {
        super.onCreate(savedInstanceState);
        new AlertDialog.Builder(this)
                .setIcon(android.R.drawable.ic_dialog_alert)
                .setTitle("闹钟")
                .setMessage("时间到")
                .setPositiveButton("关闭",new DialogInterface.OnClickListener() {
                    @Override
                    public void onClick(DialogInterface dialog,int which) {
                        finish();
```

```
            }
        }).show();
    }
}
```

5.4.2 服务（Service）的定义

1. Service 的定义

在 Android 系统开发中，Service 是一个重要的组成部分。如果某些程序不希望用户看见，那么可以将这些程序定义在 Service 中，这样就可以完成程序的后台运行（也可以在不显示界面的形式下运行）。也就是说，Service 实际上相当于是一个没有图形界面的 Activity 程序，而且当用户要执行某些操作需要进行跨进程访问时，也可以使用 Service 来完成。该类的定义如下：

```
java.lang.Object
   ↳ android.content.Context
      ↳ android.content.ContextWrapper
         ↳ android.app.Service
```

2. Service 的生命周期

与 Activity 相似，Service 也有自己的生命周期方法，如表 5-11 所示。

表 5-11 Service 类的生命周期方法列表

序 号	方 法 名 称	说 明
1	public abstract IBinder onBind（Intent intent）	设置 Activity 和 Service 之间的绑定，可由 bindService（）方法触发
2	public void onCreate（）	当 Service 在创建时被调用
3	public void onDestroy（）	Service 被销毁时调用此方法，由 stopService（）方法触发
4	public int onStartCommand（Intent intent, int flags, int startId）	启动 Service，由 startService（）方法触发
5	public boolean onUnbind（Intent intent）	取消 Activity 和 Service 之间的绑定

从表 5-11 中可以发现，onBind（）方法是一个抽象方法，在定义子类时必须重写。这个方法主要是在 Activity 和 Service 之间绑定使用。具体的绑定操作还需要 android.content.ServiceConnection 接口的支持。

3. 操作系统服务

android.content.Context 类用常量的形式绑定了所有的系统服务名称，用户使用时直接调用 getSystemService（）方法即可通过指定的服务名称取得。Context 类中定义的一些系统服务的名称如表 5-12 所示。

表 5-12 Context 类中定义的系统服务名称列表

序 号	常量名称	说　明
1	public static final String ACTIVITY_SERVICE	运行程序服务
2	public static final String ALARM_SERVICE	闹钟服务
3	public static final String AUDIO_SERVICE	音频服务
4	public static final String CLIPBOARD_SERVICE	剪贴板服务
5	public static final String LOCATION_SERVICE	位置服务
6	public static final String NOTIFICATION_SERVICE	Notification 服务
7	public static final String POWER_SERVICE	电源管理服务
8	public static final String SEARCH_SERVICE	搜索服务
9	public static final String SENSOR_SERVICE	传感器服务
10	public static final String WINDOW_SERVICE	窗口服务

例如，在"闹钟服务"应用程序中，就用 getSystemService(Context.ALARM_SERVICE) 方法获取到了闹钟管理器。

5.4.3　广播接收器（BroadcastReceiver）的定义与使用

1. BroadcastReceiver 的定义

广播是一种信息的发送机制，在 Android 手机中存在着各种各样的广播信息，如手机刚启动时的提示信息、电池不足的警告信息和来电信息等都会通过广播的形式发送给用户，而处理的形式由用户自己决定。

在 Android 中，所有的广播组件都是以类的形式存在，并且该类必须继承自 BroadcastReceiver，定义之后还必须要注册。该类的定义如下：

```
java.lang.Object
    ↳ android.content.BroadcastReceiver
```

BroadcastReceiver 类中的 public abstract void onReceive(Context context, Intent intent) 方法必须要重写。在定义自己的广播类时，可以参考以下代码：

```java
public class AlarmReceiver extends BroadcastReceiver {
    @Override
    public void onReceive(Context context, Intent intent) {
        ...
    }
}
```

在 AndroidManifest.xml 中注册时，可以参考以下代码：

```xml
<receiver
    android:name="AlarmReceiver"
    android:enabled="true" >
    <intent-filter>
        <action android:name="com.book.action.set_alarm"/>
    </intent-filter>
</receiver>
```

注册广播接收器时，需要用 <receiver> 节点表示广播组件，并且在 "android:name" 属性中指定广播程序的处理类，上述代码中处理类为 "AlarmReceiver"。为了让广播处于启用状态需要设置 "android:enabled = "true""。<intent-filter> 节点表示只有执行指定的 Action 时才会进行广播，上述代码中的 action 名称为 "com.book.action.set_alarm"。

2. PendingIntent 的定义

PendingIntent 表示当前的 Activity 不立即使用此 Intent 进行处理，而将此 Intent 封闭后传递给其他 Activity 程序，其他 Activity 程序在需要使用此 Intent 时才进行操作。此类的定义如下：

```
java.lang.Object
  ↳ android.app.PendingIntent
```

从定义中可以看到 PendingIntent 与 Intent 类之间没有任何继承关系，所以这两个类表示不同的 Intent 操作。

在 "闹钟服务" 中用到了 PendingIntent 对象，这意味着在需要时才会执行 PendingIntent 所包裹的 Intent 对象，而通过该 Intent 可以跳转到一个指定的闹钟处理程序中。

参考代码如下：

```
Intent i = new Intent(MainActivity.this, AlarmReceiver.class);
i.setAction(ACTION_ALARM);
sender = PendingIntent.getBroadcast(
    MainActivity.this, 0, i, PendingIntent.FLAG_UPDATE_CURRENT);
```

3. BroadcastReceiver 的使用步骤

使用广播接收器类时，首先要定义一个继承于 BroadcastReceiver 的子类，并在其中重写 onReceive 方法，用以实现在接收到广播之后的具体处理操作；其次，在 AndroidManifest.xml 中注册广播接收器；第三，调用 Context 类的 sendBroadcast() 方法触发广播组件。

5.4.4 四大组件之间的关系

在 Android 应用程序中，Activity、BroadcastReceiver、Service 和负责数据共享的 Content Provider 共同构成了 Android 应用程序的四大组件。除 Content Provider 之外，其余三大组件的功能及其间的关系如下：Activity 负责显示用户界面，好像是应用程序的眼睛；BroadcastReceiver 负责接收应用程序的广播，好像是应用程序的耳朵；Service 负责执行具体的操作，好像是应用程序的手。开发者可以灵活运用这些组件实现应用程序功能。

5.5 动手实践5：掌上电子邮件

5.5.1 功能要求

1. 应用程序界面

应用程序的界面如图 5-6 所示。

2. 功能描述

当用户输入合法的收件人地址和发件人地址后，即可将设定主题和内容的邮件发送出

图 5-6 "掌上电子邮件"运行效果图

去。如果用户输入的收件人地址或者发件人地址不符合邮件地址格式，则弹出如图 5-6b 所示的提示信息。

5.5.2 操作提示

1. 完成应用程序界面

应用程序中使用到了 TextView、EditText 和 Button 组件，用嵌套的线性布局或表格布局即可完成首界面设计。

2. 定义邮件地址的正则表达式

邮件地址的正则表达式为字符串类型，其值如下：

> parent = "^[a-zA-Z][\\w\\.-]*[a-zA-Z0-9]@[a-zA-Z0-9][\\w\\.-]*[a-zA-Z0-9]\\.[a-zA-Z][a-zA-Z\\.]*[a-zA-Z]$";

3. 完成发送邮件功能

发送邮件时，需要指定 Intent 的 Action 为 Intent.ACTION_SEND，数据类型为 "plain/text"，还要将收件人、发件人、邮件主题和邮件内容分别都附加到 Intent 中，对应的键名分别是 Intent.EXTRA_EMAIL、Intent.EXTRA_CC、Intent.EXTRA_SUBJECT、Intent.EXTRA_TEXT。

```
//查看地址是否符合格式
if(!strReceiver.matches(parent)){
    Toast.makeText(...).show();
}else if(!strSender.matches(parent)){
    Toast.makeText(...).show();
}else{//发送邮件
    Intent intent = new Intent(Intent.ACTION_SEND);
    intent.setType("plain/text");
    intent.putExtra(Intent.EXTRA_EMAIL,strReceiver);
    intent.putExtra(Intent.EXTRA_CC,strSender);
    intent.putExtra(Intent.EXTRA_SUBJECT,strTheme);
    intent.putExtra(Intent.EXTRA_TEXT,strMessage);
    startActivity(Intent.createChooser(intent,getResources().getString(R.string.start)));
}
```

第6章 Android 中的数据存取

在实际的应用程序开发中,数据的保存与读取是最为重要的操作。Android 也提供了专门的数据操作,用户可以利用文件或者是系统本身提供的数据库完成这种操作。在 Android 系统中共有 5 种数据存储方式:SharedPreferences 存储、文件存储、数据库存储、ContentProvider 存储和网络存储。ContentProvider 在第 5.3 节"实例 3:快速联系"中已有讲述,本章将重点讲述前 3 种存储方式。

6.1 实例 1:保存偏好设置

6.1.1 功能要求与操作步骤

1. 功能要求

完成如图 6-1 所示的应用程序。要求能够:

① 应用程序的首界面如图 6-1a 所示,单击"设置"选项菜单可以弹出图 6-1b 所示的设置界面。

② 在图 6-1b 所示的设置界面中可以设置文字的颜色和字号,效果分别如图 6-1c 和图 6-1d 所示。当选择了相应的颜色和字号再回到主界面时,文字格式发生相应改变。图 6-1e 是用户在选择了"深蓝色"和"大号字"后的效果图。

③ 当用户在设置界面中勾选了"恢复默认设置"后,"颜色"和"字号"设置项不能再使用,文字颜色恢复为黑色,文字大小恢复到 20sp。

2. 操作步骤

(1) 创建项目

参考如下输入信息创建 Android 项目:

```
Application Name:06_TestSharedPref
Project Name:06_TestSharedPref
Package Name:com.book.testsharedpref
Activity Name:MainActivity
Layout Name:activity_main
```

(2) 定义所需字符串资源

```
<?xml version="1.0" encoding="utf-8"?>
<resources>
<string name="app_name">06_TestSharedPref</string>
```

```xml
<string name = "action_settings" >设置</string>
    <string name = "content" >
        眼睛瞪得像铜铃\n
            ...
    </string>
</resources>
```

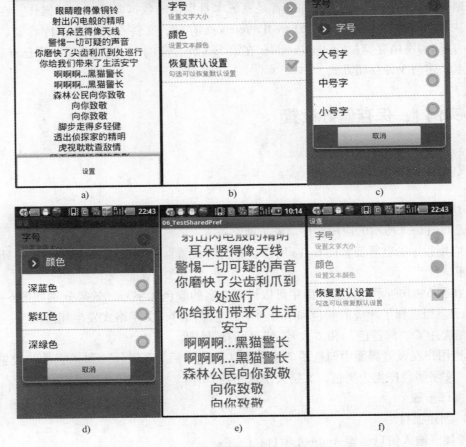

图6-1 "保存偏好设置"运行效果图

(3) 定义数组资源

设置列表所需的列表项及列表项对应的数值。

```xml
<?xml version = "1.0" encoding = "utf-8"?>
<resources>
<string-array name = "size">
<item>大号字</item>
<item>中号字</item>
<item>小号字</item>
```

```xml
</string-array>
<string-array name="color">
<item>深蓝色</item>
<item>紫红色</item>
<item>深绿色</item>
</string-array>
<string-array name="size_value">
<item>30</item>
<item>20</item>
<item>10</item>
</string-array>
<string-array name="color_value">
<item>1</item>
<item>2</item>
<item>3</item>
</string-array>
</resources>
```

(4) 定义菜单资源

主界面中有选项菜单,存放在 res/menu/main.xml 中。参考代码如下:

```xml
<menu xmlns:android="http://schemas.android.com/apk/res/android">
<item
        android:id="@+id/action_settings"
        android:title="@string/action_settings"/>
</menu>
```

(5) 定义主界面布局 activity_main

从图 6-1a 可看出,本应用主界面比较简单。参考代码如下:

```xml
<RelativeLayout xmlns:android="http://schemas.android.com/apk/res/android"
    xmlns:tools="http://schemas.android.com/tools"
    android:layout_width="match_parent"
    android:layout_height="match_parent">
    <ScrollView
    android:layout_height="match_parent"
    android:layout_width="match_parent">
<TextView
        android:id="@+id/tv"
        android:layout_width="match_parent"
        android:layout_height="match_parent"
        android:text="@string/content"
        android:gravity="center"
        android:textSize="20sp"/>
    </ScrollView>
</RelativeLayout>
```

(6) 定义设置界面 setting.xml

设置界面 setting.xml 存放在 res/xml/目录下。该界面是用于设置的界面,定义界面时可以设置根元素为"PreferenceScreen"。与其他布局不同的是在使用时不能用setContentView()方法设置Activity的布局,而是应该将此布局对应的Activity继承于PreferenceActivity,这样就可以通过调用addPreferencesFromResource()方法获得到该界面。这样做的目的在于使设置界面与Android系统的设置界面风格相一致。

在本例中,设置项共有3个,其中"颜色"和"字号"是列表形式,需要使用"ListPreference"来定义;"恢复默认设置"是复选框形式,可以使用"CheckBoxPreference"来定义。所有的设置项都有"android:key"属性,这与普通布局中组件的"android:id"属性的作用相似,用来唯一标识设置项。

本界面中各项的属性设置如表6-1所示。

表6-1 设置界面各组件属性列表

序号	组件类型	属性名称	属性值	说明
1	PreferenceScreen	xmlns:android	http://schemas.android.com/apk/res/android	设置命名域
2	ListPreference	android:key	text_size	设置唯一标识key
		android:title	字号	设置标题
		android:summary	设置文字大小	设置描述文字
		android:entries	@array/size	设置列表项
		android:entryValues	@array/size_value	设置列表项对应的值
3	ListPreference	android:key	text_color	设置唯一标识key
		android:title	颜色	设置标题
		android:summary	设置文本颜色	设置描述文字
		android:entries	@array/color	设置列表项
		android:entryValues	@array/color_value	设置列表项对应的值
4	CheckBoxPreference	android:key	default_setting	设置唯一标识key
		android:title	恢复默认设置	设置标题
		android:summary	勾选可以恢复默认设置	设置描述文字

对于ListPreference而言,需要设置好其列表项数组和列表项对应值的数组。这两个数组的类型均为<string-array>,并且数组元素个数需要保持一致。

参考代码如下:

```
<PreferenceScreen
    xmlns:android="http://schemas.android.com/apk/res/android">
<ListPreference
    android:key="text_size"
    android:title="字号"
    android:summary="设置文字大小"
    android:entries="@array/size"
    android:entryValues="@array/size_value"/>
<ListPreference
    android:key="text_color"
```

```
            android:title = "颜色"
            android:summary = "设置文本颜色"
            android:entries = "@array/color"
            android:entryValues = "@array/color_value"
            />
    <CheckBoxPreference
            android:key = "default_setting"
            android:title = "恢复默认设置"
            android:summary = "勾选可以恢复默认设置"/>
</PreferenceScreen>
```

(7) 定义 PreferSetting.java 类

这个类需要继承 PreferenceActivity，并且需要实现 OnPreferenceChangeListener 方法，用来监听设置发生改变的事件。

参考代码如下：

```java
public class PreferSetting extends PreferenceActivity implements
OnPreferenceChangeListener {
    // 定义选项的 key 和默认值
    private static final String OPT_SIZE = "text_size";
    private static final int OPT_SIZE_DEF = 20;
    private static final String OPT_COLOR = "text_color";
    private static final int OPT_COLOR_DEF = Color.BLACK;
    private static final String OPT_DEFAULT = "default_setting";
    // 声明 SharedPreferences 对象
    static SharedPreferences sp;
    // 声明 ListPreference 对象
    ListPreference lp_size, lp_color;
    // 声明 CheckBoxPreference 对象
    CheckBoxPreference chk_default;
    // 声明 SharedPreferences 编辑器对象
    Editor font_editor;
    @Override
    protected void onCreate(Bundle savedInstanceState) {
        super.onCreate(savedInstanceState);
        //设置页面布局
        addPreferencesFromResource(R.xml.settings);
        //定义相关组件
        lp_size = (ListPreference) findPreference(OPT_SIZE);
        lp_color = (ListPreference) findPreference(OPT_COLOR);
        chk_default = (CheckBoxPreference) findPreference(OPT_DEFAULT);
        //为组件添加事件监听
        lp_size.setOnPreferenceChangeListener(this);
        lp_color.setOnPreferenceChangeListener(this);
        chk_default.setOnPreferenceChangeListener(this);
        //获取本应用的 SharedPreferences
        sp = this.getSharedPreferences("UserPref", 0);
```

```java
        //从 SharedPreferences 中取出"是否默认设置"的值
        if(sp.getBoolean(OPT_DEFAULT,false)){
            //如果是默认设置,则两个列表项不可用
            lp_size.setEnabled(false);
            lp_color.setEnabled(false);
        }
    }
    //实现事件过程
    @Override
    public boolean onPreferenceChange(Preference preference,Object newValue){
        //获取 SharedPreferences 编辑器对象
        font_editor = sp.edit();
        //判断改变了哪个设置项
        if(preference.getKey().equals(OPT_SIZE)){
            //取得选择的列表项对应的下标
            int index = lp_size.findIndexOfValue((String)newValue);
            //定义列表项值对应的数组 entries
            CharSequence[] entries = lp_size.getEntryValues();
            //从列表项值数组中取出用户的选项并转换为整型
            int size_value = Integer.valueOf((String)entries[index]);
            //将用户的选择写入到 SharedPreferences 编辑器
            font_editor.putInt(OPT_SIZE,size_value);
            //提交
            font_editor.commit();
            return true;
        }
        if(preference.getKey().equals(OPT_COLOR)){
            //取得选择的列表项对应的下标
            int index = lp_color.findIndexOfValue((String)newValue);
            int color_value = Color.BLACK;
            //根据下标设置具体的颜色
            switch(index){
            case 0:
                color_value = Color.BLUE;
                break;
            case 1:
                color_value = Color.MAGENTA;
                break;
            case 2:
                color_value = Color.GREEN;
                break;
            }
            //将用户的选择写入到 SharedPreferences 编辑器
            font_editor.putInt(OPT_COLOR,color_value);
            font_editor.commit();
            return true;
        }
```

```java
            if(preference.getKey().equals(OPT_DEFAULT)){
                //如果勾选了"恢复默认设置"复选框
                //设置复选框的选中状态
                chk_default.setChecked((Boolean)newValue);
                if(newValue.equals(true)){
                    //两个列表不可用
                    lp_size.setEnabled(false);
                    lp_color.setEnabled(false);
                    //写入SharedPreferences编辑器
                    font_editor.putBoolean(OPT_DEFAULT,true);
                    font_editor.putInt(OPT_SIZE,OPT_SIZE_DEF);
                    font_editor.putInt(OPT_COLOR,OPT_COLOR_DEF);
                    font_editor.commit();
                }else{
                    //两个列表可用
                    lp_size.setEnabled(true);
                    lp_color.setEnabled(true);
                    //写入SharedPreferences编辑器
                    font_editor.putBoolean(OPT_DEFAULT,false);
                    font_editor.commit();
                }
            }
            return false;
        }
    }
```

（8）完成 MainActivity 类的定义

这个类的主要功能是使用 SharedPreferences 中存储的用户偏好设置更改文本显示框中显示文本的字体格式。

参考代码如下：

```java
public class MainActivity extends Activity {
    private static final String OPT_SIZE = "text_size";
    private static final int OPT_SIZE_DEF = 20;
    private static final String OPT_COLOR = "text_color";
    private static final int OPT_COLOR_DEF = Color.BLACK;
    SharedPreferences sp;
    TextView tv;
    @Override
    protected void onCreate(Bundle savedInstanceState) {
        super.onCreate(savedInstanceState);
        setContentView(R.layout.activity_main);
        tv = (TextView) findViewById(R.id.tv);
        //获取当前应用的SharedPreferences对象
        sp = this.getSharedPreferences("UserPref",0);
        //获取SharedPreferences中存储的用户设置值
        tv.setTextColor(sp.getInt(OPT_COLOR,OPT_COLOR_DEF));
```

```
            tv.setTextSize(sp.getInt(OPT_SIZE,OPT_SIZE_DEF));
    }
    @Override
    protected void onResume(){
        //重新返回到主界面时,字体格式根据用户选择发生改变
        tv.setTextColor(sp.getInt(OPT_COLOR,OPT_COLOR_DEF));
        tv.setTextSize(sp.getInt(OPT_SIZE,OPT_SIZE_DEF));
        super.onResume();
    }
    @Override
    public boolean onOptionsItemSelected(MenuItem item){
        //选择了菜单项后启动设置界面
        Intent i = new Intent(MainActivity.this,PreferSetting.class);
        startActivity(i);
        return super.onOptionsItemSelected(item);
    }
    @Override
    public boolean onCreateOptionsMenu(Menu menu){
        //创建选项菜单
        getMenuInflater().inflate(R.menu.main,menu);
        return true;
    }
}
```

6.1.2 SharedPreferences 的定义与使用

1. SharedPreferences 的定义

在 Android 应用程序中,往往允许用户定制化使用 APP,如是否保存密码等。如果想要在 Android 应用中实现用户偏好设置信息的保存需要使用 SharedPreferences 完成,该类定义如下:

android.content.SharedPreferences

SharedPreferences 提供了一些基础的信息保存功能,所有的信息都是按照"(键,值)"的形式进行保存,而且所保存的数据只能是一些基本的数据类型,如字符串、整型、布尔型等。SharedPreferences 常用的方法如表 6-2 所示。

表 6-2 SharedPreferences 常用的方法说明

序号	方法名称	说明
1	public abstract SharedPreferences.Editor edit()	使其处于可编辑状态
2	public abstract boolean contains(String key)	判断 key 是否存在
3	public abstract boolean getBoolean(String key, boolean defValue)	获取 boolean 类型的数据,指定默认值为 defValue
4	public abstract float getFloat(String key, float defValue)	获取 float 类型的数据,指定默认值为 defValue
5	public abstract int getInt(String key, int defValue)	获取 int 类型的数据,指定默认值为 defValue
6	public abstract long getLong(String key, long defValue)	获取 long 类型的数据,指定默认值为 defValue
7	public abstract String getString(String key, String defValue)	获取 String 类型的数据,指定默认值为 defValue

在写入数据时，必须首先调用 edit( ) 方法才可以让其处于可编辑状态，edit( ) 方法的返回值是 SharedPreferences.Editor 接口实例。在完成写操作后，调用该接口的 commit( ) 方法提交。该接口的常用方法如表 6-3 所示。

表 6-3 SharedPreferences.Editor 常用的方法说明

序号	方法名称	说明
1	public abstract SharedPreferences.Editor clear( )	清除所有数据
2	public abstract boolean commit( )	提交更新的数据
3	public abstract SharedPreferences.Editor putBoolean(String key, boolean value)	保存 boolean 类型的数据
4	public abstract SharedPreferences.Editor putFloat(String key, float value)	保存 float 类型的数据
5	public abstract SharedPreferences.Editor putInt(String key, int value)	保存 int 类型的数据
6	public abstract SharedPreferences.Editor putLong(String key, long value)	保存 long 类型的数据
7	public abstract SharedPreferences.Editor putString(String key, String value)	保存 String 类型的数据
8	public abstract SharedPreferences.Editor remove(String key)	删除指定 key 的数据

2. 使用 SharedPreferences 存取数据

使用 SharedPreferences 存取数据的基本步骤如下。

① 调用 Context 类的 getSharedPreferences(String name, int mode) 方法，指定保存操作的文件名称（不需要指定文件名称的后缀，后缀自动设置为 .xml）和操作模式。模式值可以为 0、MODE_PRIVATE、MODE_WORLD_READABLE 和 MODE_WORLD_WRITEABLE。生成的文件被保存在应用程序包中的 shared_prefs 目录下，从 DDMS 中可以看到。参考代码如下：

```
sp = this.getSharedPreferences("UserPref",0);
```

② 调用 SharedPreferences 的 edit( ) 方法生成 SharedPreferences.Editor 实例，写入相关数据，并提交。参考代码如下：

```
font_editor = sp.edit( );
font_editor.putInt("size",size_value);
font_editor.commit( );
```

③ 调用 SharedPreferences 的 getXXX( ) 方法读取数据。参考代码如下：

```
sp.getInt("size",20);
```

6.2 实例 2：贴身账簿

6.2.1 功能要求与操作步骤

1. 功能要求

完成如图 6-2 所示的应用程序。要求能够：

① 应用程序的主界面如图 6-2a 所示。用户可以选择输入记账日期，记账方式是支出或是收入，金额以及不超过 40 字的备注。

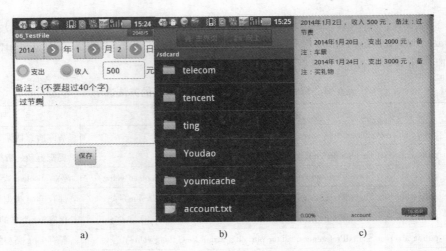

图6-2 "贴身账簿"运行效果图

② 用户输入信息后,单击"保存"按钮即可将账目信息保存在 sd 卡根目录中的 account.txt 中。从手机"我的文件"(或 DDMS/sdcard/)中即可看到如图6-2b所示的文件。

③ 打开 account.txt 文件,即可看到所有的账目记录,如图6-2c所示。

2. 操作步骤

(1) 创建项目

参考如下输入信息创建 Android 项目:

```
Application Name: 06_TestFile
Project Name: 06_TestFile
Package Name: com.book.testfile
Activity Name: MainActivity
Layout Name: activity_main
```

(2) 定义布局

本应用需要用到嵌套布局,参考代码如下:

```xml
< LinearLayout xmlns:android = "http://schemas.android.com/apk/res/android"
    android:layout_width = "fill_parent"
    android:layout_height = "fill_parent"
    android:orientation = "vertical" >
< LinearLayout
        android:layout_width = "match_parent"
        android:layout_height = "wrap_content" >
< Spinner
            android:id = "@+id/spn_year"
            android:layout_width = "0dp"
            android:layout_height = "wrap_content"
            android:layout_weight = "3"
            android:entries = "@array/year"/>
```

```xml
<TextView
    android:layout_width="wrap_content"
    android:layout_height="wrap_content"
    android:text="年"
    android:textSize="20sp"/>
<Spinner
    android:id="@+id/spn_month"
    android:layout_width="0dp"
    android:layout_height="wrap_content"
    android:layout_weight="2"
    android:entries="@array/month"/>
<TextView
    android:layout_width="wrap_content"
    android:layout_height="wrap_content"
    android:text="月"
    android:textSize="20sp"/>
<Spinner
    android:id="@+id/spn_day"
    android:layout_width="0dp"
    android:layout_height="wrap_content"
    android:layout_weight="2"
    android:entries="@array/day"/>
<TextView
    android:layout_width="wrap_content"
    android:layout_height="wrap_content"
    android:text="日"
    android:textSize="20sp"/>
</LinearLayout>
<LinearLayout
    android:layout_width="match_parent"
    android:layout_height="wrap_content" >
<RadioGroup
    android:id="@+id/rg_op"
    android:layout_width="match_parent"
    android:layout_height="wrap_content"
    android:orientation="horizontal" >
<RadioButton
    android:id="@+id/rd_out"
    android:layout_width="0dp"
    android:layout_weight="1"
    android:layout_height="wrap_content"
    android:checked="true"
    android:text="支出"/>
<RadioButton
    android:id="@+id/rd_in"
    android:layout_width="0dp"
    android:layout_weight="1"
```

```xml
                android:layout_height="wrap_content"
                android:text="收入"/>
    <EditText
                android:id="@+id/edt_money"
                android:layout_width="0dp"
                android:layout_height="wrap_content"
                android:layout_weight="1"
                android:inputType="numberDecimal"
                />
    <TextView
                android:layout_width="wrap_content"
                android:layout_height="wrap_content"
                android:text="元"
                android:textSize="20sp"/>
</RadioGroup>
</LinearLayout>
<LinearLayout
        android:layout_width="match_parent"
        android:layout_height="wrap_content"
        android:orientation="vertical" >
    <TextView
                android:layout_width="wrap_content"
                android:layout_height="wrap_content"
                android:text="备注：(不要超过40个字)"
                android:textSize="20sp"/>
    <EditText
                android:id="@+id/edt_summary"
                android:layout_width="match_parent"
                android:layout_height="wrap_content"
                android:inputType="textMultiLine"
                android:maxLength="40"
                android:lines="5"
                android:gravity="top" >
        <requestFocus />
    </EditText>
</LinearLayout>
<Button
        android:layout_width="wrap_content"
        android:layout_height="wrap_content"
        android:text="保存"
        android:layout_gravity="center"
        android:onClick="clickSave"/>
</LinearLayout>
```

（3）定义Spinner需要用的数组资源

参考代码如下：

```xml
<?xml version="1.0" encoding="utf-8"?>
```

```xml
<resources>
<string-array name="year">
<item>2013</item>
<item>2014</item>
<item>2015</item>
</string-array>
<string-array name="month">
<item>1</item>
...
<item>12</item>
</string-array>
<string-array name="day">
<item>1</item>
...
<item>31</item>
</string-array>
</resources>
```

(4) 在 MainActivity 中实现功能

MainActivity 中需要实现的功能有：第一，从各个组件中取出用户输入的信息；第二，将用户输入的信息连接成字符串；第三，将连接好的字符串存储在指定的文件中。

参考代码如下：

```java
public class MainActivity extends Activity implements OnCheckedChangeListener,OnItemSelectedListener{
    //声明组件
    Spinner spnYear,spnMonth,spnDay;
    EditText edtMoney,edtSummary;
    RadioGroup rgOp;
    //声明存储信息所用的字符串,并初始化
    String opType="支出",year="2013",month="1",day="1",record="";
    //定义文件路径及名称
    private static final String FILENAME="/sdcard/account.txt";
    @Override
    protected void onCreate(Bundle savedInstanceState){
        super.onCreate(savedInstanceState);
        setContentView(R.layout.activity_main);
        //定义组件
        spnYear=(Spinner)findViewById(R.id.spn_year);
        spnMonth=(Spinner)findViewById(R.id.spn_month);
        spnDay=(Spinner)findViewById(R.id.spn_day);
        edtMoney=(EditText)findViewById(R.id.edt_money);
        edtSummary=(EditText)findViewById(R.id.edt_summary);
        rgOp=(RadioGroup)findViewById(R.id.rg_op);
        //为组件添加监听
        spnYear.setOnItemSelectedListener(this);
        spnMonth.setOnItemSelectedListener(this);
```

```java
        spnDay.setOnItemSelectedListener(this);
        rgOp.setOnCheckedChangeListener(this);
}
public void clickSave(View v){
    //连接字符串
    record = year + "年" +
            month + "月" +
            day + "日," +
            opType + " " +
            edtMoney.getText().toString() + "元" +
            ",备注:" + edtSummary.getText().toString();
    //保存
    saveAll(record);
}
private void saveAll(String str){
    // TODO Auto-generated method stub
    File file = new File(FILENAME);
    if(!file.getParentFile().exists()){
        file.getParentFile().mkdir();
    }
    PrintStream out = null;
    try{
        out = new PrintStream(new FileOutputStream(file,true));
        out.println(str);
    }catch(Exception e){
        e.getCause();
        Toast.makeText(MainActivity.this,"保存失败,请查看SD卡是否存在",
            Toast.LENGTH_SHORT).show();
    }finally{
        if(out!=null){
            out.close();
        }
    }
    finish();
}
//实现单选按钮组的监听事件
@Override
public void onCheckedChanged(RadioGroup group,int checkedId){
    // TODO Auto-generated method stub
    switch(checkedId){
    case R.id.rd_in:
        opType = ((RadioButton)group.getChildAt(1)).getText().toString();
        break;
    case R.id.rd_out:
        opType = ((RadioButton)group.getChildAt(0)).getText().toString();
        break;
    }
```

```
            }
            //实现 Spinner 的项目选择事件
            @Override
            public void onItemSelected(AdapterView<?> adapterView,View v,int pos,
                    long id) {
                switch(adapterView.getId()){
                case R.id.spn_year:
                    year = adapterView.getItemAtPosition(pos).toString();
                    break;
                case R.id.spn_month:
                    month = adapterView.getItemAtPosition(pos).toString();
                    break;
                case R.id.spn_day:
                    day = adapterView.getItemAtPosition(pos).toString();
                    break;
                }
            }
            @Override
            public void onNothingSelected(AdapterView<?> arg0) {
            }
        }
```

(5) 在 AndroidManifest.xml 中配置应用程序

由于本程序要使用到外部设备,所以必须要为程序配置相应的权限。

参考代码如下:

```
<uses-permission android:name="android.permission.WRITE_EXTERNAL_STORAGE"/>
```

6.2.2 文件(File)的定义与使用

1. File 的定义

Android 应用程序还可以将一些需要永久保存的数据以文件(File)的形式保存在设备的 SD 卡中。File 类的定义如下:

```
java.lang.Object
    ↳ java.io.File
```

File 类是 java.io 包中唯一代表磁盘文件本身的对象。File 类定义了一些与平台无关的方法来操作文件,可以通过调用 File 类中的方法,实现文件的创建、删除以及重命名等操作。

File 常用的方法如表 6-4 所示。

表 6-4 File 常用的方法说明

序号	方法名称	说明
1	public boolean createNewFile()	创建新的空文件
2	public static File createTempFile(String prefix, String suffix, File directory)	在指定目录中创建前缀(不超过3个字符)、后缀确定的临时文件

(续)

序号	方法名称	说明
3	public static FilecreateTempFile(String prefix, String suffix)	创建临时文件，参数为前缀名（不超过3个字符）和后缀名
4	public boolean delete()	删除文件或目录，删除目录时需要保证目录内容为空
5	public boolean equals(Object obj)	比较两个文件是否是相同的对象
6	public boolean exists()	判断文件是否存在
7	public File(String path)	构造函数，根据参数指定路径构造文件对象
8	public String getAbsolutePath()	获取文件的绝对路径，android中的根目录是"/"
9	public String getParent()	获取文件的父目录路径
10	public File getParentFile()	返回从文件父目录中生成的一个新文件
11	public String getPath()	获取文件路径
12	public boolean isDirectory()	判断文件是否为目录
13	public boolean isFile()	判断是否为文件
14	public long length()	获取文件的长度（单位：字节）
15	public File[] listFiles()	获取文件目录中存在的所有文件并存到文件数组中
16	public boolean mkdir()	创建子目录

2. 使用 File 和文件输入输出流存取数据

使用 File 存取数据的基本步骤如下。

① 创建文件对象。参考代码如下：

```
File file = new File("/sdcard/mydoc.txt");
```

② 写文件时，首先需要用 FileOutputStream 创建文件输出流，并以此为参数，使用 PrintStream(OutputStream out) 构造函数创建 PrintStream 对象，然后调用 PrintStream 类的 println(String str) 方法完成写文件操作，最后关闭 PrintStream 对象。参考代码如下：

```
PrintStream out = new PrintStream(new FileOutputStream(file,true));
out.println(str);
out.close();
```

③ 读文件时，首先使用 FileReader 类构造输入流，然后使用 BufferedReader 构造带缓存的输入流，接着在循环中调用 BufferedReader 类的 readLine() 读行，并将读出的数据显示，最后关闭流对象。参考代码如下：

```
FileReader file = null;
try {
    file = new FileReader("/sdcard/mydoc.txt");
} catch (FileNotFoundException e) {
    e.printStackTrace();
}
try {
    BufferedReader br = new BufferedReader(file);
    String s = null;
```

```
            int i = 0;
            while((s = br.readLine()) != null){
                tv.setText(tv.getText() + " \n" + s);
                i++;
            }
            br.close();
            file.close();
        } catch(IOException e){
            e.printStackTrace();
        }
```

在用 File 存储数据时，需要考虑到用户要自定义保存目录，以及在 SD 卡上操作。所以需要为程序配置相应的权限。

```
<uses-permission
    android:name="android.permission.WRITE_EXTERNAL_STORAGE"
/>
```

6.3 实例3：备忘随行

6.3.1 功能要求与操作步骤

1. 功能要求

完成如图 6-3 所示的应用程序。要求能够：

① 应用程序的主界面如图 6-3a 所示。单击上方的"创建备忘"按钮即可显示图 6-3b 所示的标题为"增加数据"的界面，在此界面中写好信息后，单击下方的"保存"按钮即可将备忘保存在数据库中，然后返回主界面，同时主界面的数据更新，如图 6-3c 所示。

② 当用户单击主界面中的某列表项时，会弹出如图 6-3d 所示的标题为"修改数据"页面，用户可以针对当前记录的相关属性进行修改，单击下方的"保存"按钮就可将修改信息保存在数据库中。

③ 当用户长时间按下主界面中的某列表项时，会弹出如图 6-3e 所示的删除对话框，单击对话框中的"是"按钮即可将此列表项对应的数据删除。

2. 操作步骤

（1）创建项目

参考如下输入信息创建 Android 项目：

```
Application Name:06_TestSQL
Project Name:06_TestSQL
Package Name:com.book.testsql
Activity Name:MainActivity
Layout Name:activity_main
```

（2）定义主界面布局 activity_main.xml

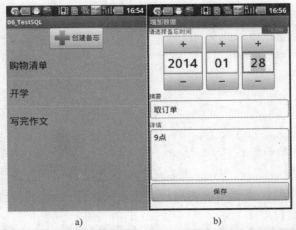

图6-3 "备忘随行"运行效果图

从图6-3a可看出主界面由一个按钮和一个ListView垂直排列而成,实现比较简单,其参考代码如下:

```
< LinearLayout xmlns:android = " http://schemas. android. com/apk/res/android"
    android:layout_width = " fill_parent"
    android:layout_height = " fill_parent"
    android:orientation = " vertical"
    android:background = " #f80" >
< Button
    android:layout_width = " wrap_content"
    android:layout_height = " wrap_content"
    android:text = " 创建备忘"
    android:onClick = " clickAdd"
    android:drawableLeft = " @drawable/add"
    android:layout_gravity = " center_horizontal" />
< ListView
    android:id = " @ + id/lv_memory"
    android:layout_width = " match_parent"
```

```
            android:layout_height = "wrap_content" >
    </ListView >
</LinearLayout >
```

(3) 定义增加数据和修改数据的界面布局 insert_modify_layout.xml

从图 6-3b 和图 6-3d 可看出，这两个界面完全相同，所以此处定义一个布局供两个功能使用。在这个布局中，使用到了 3 个起提示作用的 TextView、1 个用于输入日期的日期选择器、2 个可以输入备忘摘要和备忘详情的 EditText，这些组件垂直排列在线性布局中即可。参考代码如下：

```
< LinearLayout xmlns:android = "http://schemas.android.com/apk/res/android"
    android:layout_width = "match_parent"
    android:layout_height = "match_parent"
    android:orientation = "vertical" >
< TextView
        android:layout_width = "wrap_content"
        android:layout_height = "wrap_content"
        android:text = "请选择备忘时间"/>
< DatePicker
        android:id = "@+id/dp_date"
        android:layout_width = "match_parent"
        android:layout_height = "wrap_content"/>
< TextView
        android:layout_width = "wrap_content"
        android:layout_height = "wrap_content"
        android:text = "摘要"/>
< EditText
        android:id = "@+id/edt_sum"
        android:layout_width = "match_parent"
        android:layout_height = "wrap_content"
        android:lines = "1" >
< requestFocus />
</EditText >
< TextView
        android:layout_width = "wrap_content"
        android:layout_height = "wrap_content"
        android:text = "详情"/>
< EditText
        android:id = "@+id/edt_content"
        android:layout_width = "match_parent"
        android:layout_height = "wrap_content"
        android:lines = "5"
        android:gravity = "top"/>
< Button
        android:layout_width = "match_parent"
        android:layout_height = "wrap_content"
        android:onClick = "clickSave"
```

```
            android:text = "保存"/>
</LinearLayout >
```

(4) 定义数据库常量类 DBConstants.java

为了便于统一访问数据库，数据库名称、数据表的名称以及数据表中各列的名称全都在该类中定义，参考代码如下：

```
import android.provider.BaseColumns;
public class DBConstants implements BaseColumns {
    //定义数据库的名称
    public static final String DB_NAME = "memory.db";
    //定义表的名称
    public static final String TBL_NAME = "memory_table";
    //定义表中各列的名称
    public static final String YEAR = "year";
    public static final String MONTH = "month";
    public static final String DAY = "day";
    public static final String SUMMARY = "summary";
    public static final String CONTENT = "content";
}
```

(5) 定义数据库辅助类 DBHelper.java

使用数据库辅助类可以用来管理数据库的创建和版本更新。在 DBHelper 中还可以为数据库设置增加、删除、修改、查询数据的功能。

定义 DBHelper 辅助类时，需要使之继承于 SQLiteOpenHelper 类，并实现其中的两个抽象方法 public void onCreate(SQLiteDatabase db) 和 public void onUpgrade(SQLiteDatabase db, int oldVersion, int newVersion)，同时定义一个构造函数。

参考代码如下：

```
import static com.book.testsql.DBConstants.*;
...
public class DBHelper extends SQLiteOpenHelper {
    //创建表的语句
    private static final String CREATE_TBL = "create table"
        + TBL_NAME + "(" + _ID + " integer primary key autoincrement,"
        + YEAR + " integer," + MONTH + " integer," + DAY + " integer,"
        + SUMMARY + " text," + CONTENT + " text);";
    //更新数据的语句
    private static final String MODIFY_SQL = "UPDATE" + TBL_NAME + "SET" +
        YEAR + " = ?," + MONTH + " = ?," + DAY + " = ?," +
        SUMMARY + " = ?," + CONTENT + " = ?" + "WHERE" + _ID + " = ?;";
    //构造函数
    public DBHelper(Context context, String name, CursorFactory factory, int version) {
        super(context, name, factory, version);
    }
    //创建表
    @Override
```

```java
public void onCreate(SQLiteDatabase db) {
    db.execSQL(CREATE_TBL);
}
//插入数据
public void insert(ContentValues cv) {
    //获取数据库对象
    SQLiteDatabase db = this.getWritableDatabase();
    //插入数据
    db.insert(TBL_NAME, null, cv);
    db.close();
}
//查询全部数据
public Cursor query() {
    //获取数据库对象
    SQLiteDatabase db = this.getReadableDatabase();
    //查询
    Cursor c = db.query(TBL_NAME, null, null, null, null, null, null);
    return c;
}
//查询指定数据
public Cursor query(int id) {
    //获取数据库对象
    SQLiteDatabase db = this.getReadableDatabase();
    String[] selArgs = new String[]{String.valueOf(id)};
    //查询
    Cursor c = db.query(TBL_NAME, null, _ID + " = ?", selArgs, null, null, null);
    return c;
}
//修改数据
public void modify(int y, int m, int d, String title, String detail, int id) {
    //获取数据库对象
    SQLiteDatabase db = this.getWritableDatabase();
    Object args[] = new Object[]{y, m, d, title, detail, id};
    db.execSQL(MODIFY_SQL, args);
    db.close();
}
//删除指定数据
public void delete(int id) {
    SQLiteDatabase db = this.getWritableDatabase();
    String[] whereArgs = new String[]{String.valueOf(id)};
    db.delete(TBL_NAME, _ID + " = ?", whereArgs);
}
@Override
public void onUpgrade(SQLiteDatabase db, int oldVersion, int newVersion) {
}
}
```

在上述代码中增加、查询、删除数据使用的是 SQLiteDatabase 类的 insert(String table,

String nullColumnHack,ContentValues values)、query(String table,String[ ]columns,String selection,String[ ] selectionArgs,String groupBy,String having,String orderBy)和delete(String table, String whereClause,String[ ]whereArgs)方法。使用这些方法时，只需要传递相关的参数即可调用；在完成修改功能时使用了 SQLiteDatabase 类的 execSQL(String sql)方法，其参数是标准的 SQL 语句。修改功能也可以使用 SQLiteDatabase 类的 update(String table, ContentValues values, String whereClause, String[ ] whereArgs) 实现。

（6）实现 MainActivity 中的功能

主界面中主要有两个功能需要实现：第一，将数据库中所有记录的摘要显示在 ListView 中；第二，为 ListView 添加 OnItemClickListener 和 OnItemLongClickListener 事件，并使之能够响应相关操作。

参考代码如下：

```java
public class MainActivity extends Activity implements OnItemClickListener,OnItemLongClickListener {
    DBHelper helper = null;
    ListView lv;
    Cursor cursor;
    //定义两个 List 对象,分别用来存放列表中显示的数据和列表数据对应的 ID 字段值
    private List < String > currentData = new ArrayList < String > ( );
    private List < Integer > currentDataId = new ArrayList < Integer > ( );
    @Override
    protected void onCreate(Bundle savedInstanceState) {
        super.onCreate(savedInstanceState);
        setContentView(R.layout.activity_main);
        lv = (ListView) findViewById(R.id.lv_memory);
        lv.setOnItemClickListener(this);
        lv.setOnItemLongClickListener(this);
        //创建数据库辅助类的对象
        helper = new DBHelper(this,DB_NAME,null,1);
        //加载数据并在列表中显示
        loadData();
    }
    //当继续时,重新加载数据并显示
    @Override
    protected void onResume() {
        super.onResume();
        loadData();
    }
    //加载数据库的数据
    private void loadData() {
        //将原有的列表数据清除
        currentData.clear();
        currentDataId.clear();
        //执行查询
        cursor = helper.query();
        //当查询结果不为空时
        if(cursor! = null)
```

```java
            int id;
            String summary;
            //在查询结果中循环
            for(cursor.moveToFirst();! cursor.isAfterLast();cursor.moveToNext()){
                //取出每条记录对应的摘要和id
                summary = cursor.getString(cursor.getColumnIndex(SUMMARY));
                id = cursor.getInt(cursor.getColumnIndex(_ID));
                //将摘要和id分别添加到currentData和currentDataId列表中
                currentData.add(summary);
                currentDataId.add(id);
            }
            //用"摘要"currentData生成适配器
            ArrayAdapter<String> aa = new ArrayAdapter<String>(
                MainActivity.this,
                android.R.layout.simple_list_item_1,
                currentData);
            //列表视图使用适配器
            lv.setAdapter(aa);
        }
    }
    @Override
    protected void onDestroy(){
        //关闭数据库
        helper.close();
        super.onDestroy();
    }
    public void clickAdd(View v){
        //打开插入数据的界面
        Intent i = new Intent(MainActivity.this,InsertActivity.class);
        startActivity(i);
    }
    //单击列表项打开修改界面
    @Override
    public void onItemClick(AdapterView<?> adapterView,View v,int pos,long viewId){
        //取得选择的列表项id
        int id = currentDataId.get(pos);
        //查询_ID为指定id的数据记录
        Cursor cursor = helper.query(id);
        if(cursor! = null){
            //将游标指向第一条记录
            cursor.moveToFirst();
            //取出列表项所包含的全部字段值
            int y = cursor.getInt(cursor.getColumnIndex(YEAR));
            int m = cursor.getInt(cursor.getColumnIndex(MONTH));
            int d = cursor.getInt(cursor.getColumnIndex(DAY));
            String summary = cursor.getString(cursor.getColumnIndex(SUMMARY));
            String content = cursor.getString(cursor.getColumnIndex(CONTENT));
```

```
            //将数据存放到 Bundle 中
            Bundle data = new Bundle();
            data.putInt("id",id);
            data.putInt("y",y);
            data.putInt("m",m);
            data.putInt("d",d);
            data.putString("summary",summary);
            data.putString("content",content);
            //定义意图
            Intent i = new Intent(MainActivity.this,ModifyActivity.class);
            //携带数据
            i.putExtras(data);
            //打开修改数据的界面
            startActivity(i);
        }
    }
    //长时间按下列表项弹出删除对话框
    @Override
    public boolean onItemLongClick(AdapterView<?> arg0,View arg1,int pos,long arg3){
        final int id = currentDataId.get(pos);
        AlertDialog ad = new AlertDialog.Builder(MainActivity.this)
            .setTitle("确定删除此项?")
            .setMessage("您确定要删除这条备忘吗?")
            .setPositiveButton("是",new OnClickListener(){
                @Override
                public void onClick(DialogInterface arg0,int arg1){
                    // TODO Auto-generated method stub
                    helper.delete(id);
                    loadData();
                }
            })
            .setNegativeButton("否",new OnClickListener(){
                @Override
                public void onClick(DialogInterface dialog,int which){
                    // TODO Auto-generated method stub
                }
            })
            .create();
        ad.show();
        return false;
    }
}
```

(7) 插入数据的实现 InsertActivity.java

插入操作时，可以用 ContentValues 将需要插入的数据封装，然后再用数据库辅助类的 insert(ContentValues cv)方法将数据添加到数据库。

参考代码如下：

```java
public class InsertActivity extends Activity {
    DatePicker dp;
    EditText edtSum,edtContent;
    int year,month,day;
    ContentValues cv = new ContentValues();
    @Override
    protected void onCreate(Bundle savedInstanceState) {
        // TODO Auto-generated method stub
        super.onCreate(savedInstanceState);
        setContentView(R.layout.insert_modify_layout);
        dp = (DatePicker)findViewById(R.id.dp_date);
        edtSum = (EditText)findViewById(R.id.edt_sum);
        edtContent = (EditText)findViewById(R.id.edt_content);
    }
    public void clickSave(View v) {
        year = dp.getYear();
        month = dp.getMonth();
        day = dp.getDayOfMonth();
        cv.put(YEAR,year);
        cv.put(MONTH,month);
        cv.put(DAY,day);
        cv.put(SUMMARY,edtSum.getText().toString());
        cv.put(CONTENT,edtContent.getText().toString());
        DBHelper db = new DBHelper(this.getApplicationContext(),DB_NAME,null,1);
        db.insert(cv);
        db.close();
        finish();
    }
}
```

（8）修改数据的实现 ModifyActivity.java

在修改数据时，首先需要使用 Intent 将原有数据传递到修改界面中，然后再将修改后的数据通过调用数据库辅助类的 modify(int y, int m, int d, String title, String detail, int id) 方法写入数据库。参考代码如下：

```java
public class ModifyActivity extends Activity{
    DatePicker dp;
    EditText edtSum,edtContent;
    int year,month,day;
    ContentValues cv = new ContentValues();
    @Override
    protected void onCreate(Bundle savedInstanceState) {
        super.onCreate(savedInstanceState);
        this.setContentView(R.layout.insert_modify_layout);
        //取出列表项的所有字段值
        Intent i = getIntent();
        int y = i.getIntExtra("y",1900);
```

```
            int m = i.getIntExtra("m",1);
            int d = i.getIntExtra("d",1);
            String summary = i.getStringExtra("summary");
            String content = i.getStringExtra("content");
            //找组件并赋初始值
            dp = (DatePicker) findViewById(R.id.dp_date);
            dp.updateDate(y,m,d);
            edtSum = (EditText) findViewById(R.id.edt_sum);
            edtSum.setText(summary);
            edtContent = (EditText) findViewById(R.id.edt_content);
            edtContent.setText(content);
        }
        //按钮的单击事件
        public void clickSave(View v){
            //取得组件中的修改值
            year = dp.getYear();
            month = dp.getMonth();
            day = dp.getDayOfMonth();
            int id = getIntent().getExtras().getInt("id");
            DBHelper db = new DBHelper(this.getApplicationContext(),DB_NAME,null,1);
            //调用修改数据的方法 modify
            db.modify(year,month,day,
                    edtSum.getText().toString(),
                    edtContent.getText().toString(),
                    id);
            db.close();
            //关闭当前界面
            finish();
        }
}
```

(9) 在 AndroidManifest.xml 中配置 Activity

为了便于统一访问数据库,数据库名称、数据表的名称以及数据表中各列的名称全都在该类中定义,参考代码如下:

```
<activity
    android:name="com.book.testsql.MainActivity"
    android:label="@string/app_name" >
<intent-filter>
<action android:name="android.intent.action.MAIN"/>
<category android:name="android.intent.category.LAUNCHER"/>
</intent-filter>
</activity>
<activity
    android:name="InsertActivity"
    android:label="增加数据"/>
<activity
```

```
android:name = "ModifyActivity"
android:label = "修改数据"/>
```

6.3.2 SQLite 数据库的基本使用方法

1. SQLite 简介

SQLite 是一个轻量级的、嵌入式的关系型数据库，主要针对各种嵌入式设备而设计。由于其本身占用的存储空间较小，目前已经在 Android 操作系统中广泛使用。在 SQLite 数据库中可以方便地使用 SQL 语句或 SQLiteDatabase 类的相关方法实现数据的增加、修改、删除和查询等操作。

2. SQLiteDatabase 的定义

在 Android 中，每个 SQLiteDatabase 实例都代表了一个 SQLite 数据库的操作，通过 SQLiteDatabase 类可以执行 SQL 语句，以完成对数据的增加、删除、修改、查询等常见操作。类的定义如下：

```
java.lang.Object
  ↳ android.database.sqlite.SQLiteClosable
    ↳ android.database.sqlite.SQLiteDatabase
```

SQLiteDatabase 类的常用方法如表 6-5 所示。

表 6-5 SQLiteDatabase 类常用的方法说明

序号	方法名称	说明
1	public int delete (String table, String whereClause, String[] whereArgs)	删除数据，参数分别是表名、where 子句和参数
2	public void execSQL (String sql)	执行 SQL 语句
3	public long insert (String table, String nullColumnHack, ContentValues values)	插入数据、参数分别是表名、传入数据的列名和插入的数据
4	public int update (String table, ContentValues values, String whereClause, String[] whereArgs)	修改数据，参数分别是表名、更新的数据、where 子句和 where 参数
5	public void close()	继承于 SQLiteClosable 类的方法，用来关闭数据库
6	public Cursor query (String table, String[] columns, String selection, String[] selectionArgs, String groupBy, String having, String orderBy)	查询数据，参数分别是表名，查询的列名，where 子句，where 条件参数，分组、过滤和排序字段

3. SQLiteOpenHelper 的定义与使用

SQLiteOpenHelper 类是一个抽象类，它可以帮助用户进行数据库操作，使用时需要定义其子类，并在子类中重写相应的抽象方法。该类的定义如下：

```
java.lang.Object
  ↳ android.database.sqlite.SQLiteOpenHelper
```

SQLiteOpenHelper 类的常用方法如表 6-6 所示。

表 6-6 SQLiteOpenHelper 类常用的方法说明

序号	方法名称	说明
1	public synchronized void close()	关闭数据库
2	public SQLiteDatabase getReadableDatabase()	获取只读的数据库
3	public SQLiteDatabase getWritableDatabase()	获取可写的数据库
4	public abstract void onCreate(SQLiteDatabase db)	创建数据表，此方法不是在实例化 SQLiteOpenHelper 类时调用，而是在对象调用了 getReadableDatabase() 或 getWritableDatabase() 后会被调用
5	public abstract void onUpgrade(SQLiteDatabase db, int oldVersion, int newVersion)	更新数据表，当数据库需要进行升级时会调用此方法，一般可以在此方法中将数据表删除，并且在删除表之后往往会调用 onCreate() 方法重新创建新的数据表

例如，定义 DBHelper 辅助类之后，在 onCreate() 方法中执行创建数据表的语句，参考代码如下：

```
public void onCreate(SQLiteDatabase db) {
    String CREATE_TBL = "create table"
        + TBL_NAME + "(" + _ID + " integer primary key autoincrement," + YEAR
        + " integer," + MONTH + " integer," + DAY + " integer," + SUMMARY
        + " text," + CONTENT + " text);";
    db.execSQL(CREATE_TBL);
}
```

4. ContentValues 的定义与使用

在 SQLiteDatabase 类中提供的 insert()、update()、delete()、query() 方法均需要使用 ContentValues 类进行封装。该类的定义如下：

java.lang.Object
↳ android.content.ContentValues

ContentValues 类的常用方法如表 6-7 所示。

表 6-7 ContentValues 类常用的方法说明

序号	方法名称	说明
1	public ContentValues()	构造函数，创建 ContentValues 对象
2	public void clear()	清空全部数据
3	public Object get(String key)	根据 key 获取数据
4	public void put(String key, 类型 value)	设置 key 键对应的数据 value
5	public int size()	返回保存数据的个数

例如，创建 DBHelper 辅助类中的 insert 方法时，需要给定 ContentValues 类型的参数，然后调用 SQLiteDatabase 的 insert(String table, String nullColumnHack, ContentValues values) 方法执行插入操作。参考代码如下：

```
public void insert(ContentValues cv) {
    SQLiteDatabase db = this.getWritableDatabase();
```

```
db.insert(TBL_NAME,null,cv);
db.close();
}
```

在使用时,创建好 ContentValues 对象,将相应的值保存在其中,再调用 DBHelper 辅助类的 insert(ContentValues cv)即可。参考代码如下:

```
ContentValues cv = new ContentValues();
cv.put(YEAR,year);
cv.put(MONTH,month);
DBHelper db = new DBHelper(this.getApplicationContext(),DB_NAME,null,1);
db.insert(cv);
```

6.4 动手实践6:查账单

6.4.1 功能要求

1. 应用程序界面

应用程序的界面如图 6-4 所示。

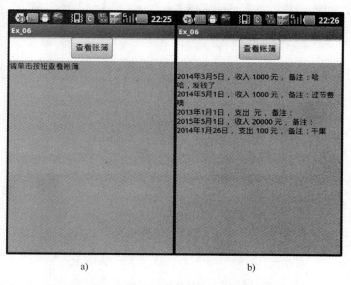

图 6-4 "查账单"运行效果图

2. 功能描述

用户单击图 6-4a 中的"查看账簿"按钮后即可在界面下方的文本区显示实例 2 "贴身账簿"中生成文件的全部内容。

6.4.2 操作提示

1. 完成应用程序首页面

本界面非常简单,需要注意的是,如果文件内容较多,文本区不能在屏幕范围内显示全部内容时,需要增加滚动条。

2. 在 Activity 中读取文件

读取文件的要领参考第 6.2.2 节介绍的内容,核心代码如下:

```java
FileReader file = null;
try {
        file = new FileReader(
                Environment.getExternalStorageDirectory().toString()
                + File.separator
                + FILENAME);
} catch (FileNotFoundException e) {
    e.printStackTrace();}
try {
        BufferedReader br = new BufferedReader(file);
        String s = null;
        int i = 0;
        while((s = br.readLine()) != null) {
            tv.setText(tv.getText() + "\n" + s);
            i++;      }
        br.close();
        file.close();
} catch (IOException e) {
        e.printStackTrace();}
```

第7章 Android 中的媒体处理

在 Android 手机应用中，图形、音频、视频等媒体的使用屡见不鲜，绘制图形、播放音乐、修改图片等功能在各种型号的手机上几乎随处可见。Android 系统如何处理这些媒体资源是本章讨论的主要内容。

7.1 实例1：绘制五星红旗

7.1.1 功能要求与操作步骤

1. 功能要求

完成"绘制五星红旗"的应用程序。要求能够：

① 根据手机屏幕方向，在手机屏幕上绘制符合比例的五星红旗。

② 在绘制时，注意旗面为红色，长宽比例为 3∶2；五颗星为黄色。

五星红旗的具体绘制方法如下：

为便于确定五星之位置，可以先将旗面对分为四个相等的长方形，然后将左上方的长方形上下划为十等分，左右划为十五等分。

大五角星的中心点在左上长方形的上五下五、左五右十之处。以此点为圆心，以三等分为半径作一圆。在此圆周上，定出五个等距离的点，其一点须位于圆的正上方。然后将此五点中各相隔的两点相连，使各成一直线。此五直线所构成的外轮廓线，就是大五角星，五角星的一个角尖正向上方。

四颗小五角星的中心点，第1点在该长方形上二下八、左十右五之处，第2点在上四下六、左十二右三之处，第3点在上七下三、左十二右三之处，第4点在上九下一、左十右五之处。其画法为：以以上4点为圆心，各以一等分为半径，分别作4个圆。在每个圆上各定出5个等距离的点，其中均须各有一点位于大五角星中心点与以上4个圆心的各联结线上。然后用构成大五角星的同样方法，构成小五角星。这4颗小五角星均各有一个角尖正对大五角星的中心点。

2. 操作步骤

（1）创建项目

参考如下输入信息创建 Android 项目：

> Application Name：07_TestDrawSim
> Project Name：07_TestDrawSim
> Package Name：com. book. testdrawsim
> Activity Name：MainActivity

211

（2）学习绘制正五角星的方法

画半径为 r（下面假设 r=1）的正圆内接正五角星的思路如下：

如图 7-1 所示，利用正五角星相邻两条边对应圆心角相同的规则，从 BD 边开始绘制，逆时针方向依次画出 10 条边。

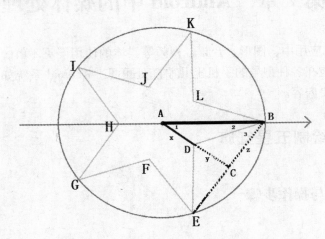

图 7-1　五角星的画法示意图

绘制 BD 边时，需要确定起点和终点的坐标位置。起点 B 是正五角星凸出的顶点，其坐标容易测算为（1，0）。而 D 点是正五角星凹进去的顶点，其坐标应为（x·cos∠1，x·sin∠1）。

依此类推，可得其余各点的坐标依次如下。

E：(sin2∠1，cos2∠1)

F：(x·cos3∠1，x·sin3∠1)

G：(sin4∠1，cos4∠1)

H：(x·cos5∠1，x·sin4∠1)

I：(sin6∠1，cos6∠1)

J：(x·cos7∠1，x·sin7∠1)

K：(sin8∠1，cos8∠1)

L：(x·cos9∠1，x·sin9∠1)

凹进去的顶点构成的正圆半径 x 的计算过程如下：

∵ 正五角星中 ∠1 = π/5，∠2 = (π/5)/2 = π/10

∴ ∠3 = (π/2 − ∠1) − ∠2 = (π/2 − π/5) − π/10 = π/5

又∵ 在直角三角形 ABC 中，z = r·sin∠1 = sinπ/5

且在直角三角形 BCD 中，y/z = tan∠3 = sin∠3/cos∠3 = (sinπ/5)/(cosπ/5)

∴ y = z·(sinπ/5)/(cosπ/5) = (sinπ/5)·(sinπ/5)/(cosπ/5)

又∵ 在直角三角形 ABC 中，x + y = r·cos∠1 = cosπ/5

∴ x = cosπ/5 − (sinπ/5)·(sinπ/5)/(cosπ/5) = (cos 2π/5)/(cosπ/5)

将 x 代入到各点坐标表达式，即可得出各点坐标值。

(3) 自定义视图 FlagView

在绘制图形时，需要自定义视图。在定义视图时，要继承 View 类，并实现自定义类的构造函数和 onDraw(Canvas canvas) 方法。

在 onDraw(Canvas canvas) 方法中，需要定义画笔 Paint 对象，并根据需要设置 Paint 对象的颜色、填充模式、是否抗锯齿以及线宽等属性。绘制图形时，可以使用画布 Canvas 类的一系列方法，如 drawLine() 绘制线段、drawCircle() 绘制圆形等。

参考代码如下：

```java
public class FlagView extends View {
    private Path mBigStarPath = new Path();
    private Path [] mSmallStarsPath;
    public FlagView(Context context) {
        super(context);
    }
    @Override
    protected void onDraw(Canvas canvas) {
        super.onDraw(canvas);
        //取得屏幕的宽,计算背景的大小
        float width = canvas.getWidth();
        float height = canvas.getHeight();
        //声明旗面长和宽
        float bgW,bgH;
        //根据屏幕方向确定长、宽
        if(width > height) {
            bgH = height;
            bgW = bgH * 3/2;
        } else {
            bgW = width;
            bgH = bgW/3 * 2;
        }
        //创建绘制旗面的画笔
        Paint bgPaint = new Paint();
        //设置旗面的画笔颜色
        bgPaint.setColor(Color.RED);
        //设置抗锯齿模式
        bgPaint.setAntiAlias(true);
        bgPaint.setStyle(Style.FILL);
        //绘制红旗背景
        canvas.drawRect(0,0,bgW,bgH,bgPaint);
        //设置绘制五角星的画笔
        Paint mStarPaint = new Paint();
        mStarPaint.setColor(Color.YELLOW);
        mStarPaint.setStyle(Style.FILL);
        mStarPaint.setStrokeWidth(3);
```

```java
mStarPaint.setAntiAlias(true);
//设置大五角星绘制路径
mBigStarPath = createStarPath(new PointF(bgW/6,bgH/4),bgW/10,-90);
//绘制大五角星
canvas.drawPath(mBigStarPath,mStarPaint);

//声明小五角星路径
mSmallStarsPath = new Path[4];
//设置小五角星对应圆的半径
float cellWidth = bgW/30;
mSmallStarsPath[0] = createStarPath(
    new PointF(cellWidth * 10,cellWidth * 2),
    cellWidth,
    (float)(Math.atan2(3,-5)/Math.PI * 180));
mSmallStarsPath[1] = createStarPath(
    new PointF(cellWidth * 12,cellWidth * 4),
    cellWidth,
    (float)(Math.atan2(1,-7)/Math.PI * 180));
mSmallStarsPath[2] = createStarPath(
    new PointF(cellWidth * 12,cellWidth * 7),
    cellWidth,
    (float)(Math.atan2(-2,7)/Math.PI * 180));
mSmallStarsPath[3] = createStarPath(
    new PointF(cellWidth * 10,cellWidth * 9),
    cellWidth,
    (float)(Math.atan2(-4,5)/Math.PI * 180 - 30));
//绘制小五角星
for(int i = 0;i < mSmallStarsPath.length;i++){
    canvas.drawPath(mSmallStarsPath[i],mStarPaint);
}
}
private Path createStarPath(PointF centerPointF,float radius, float rotate){
    //五角星中两个相邻顶点与中心点的连线之间的偏移角度为:(2/5)*PI(弧度)
    //五角星中凸出的顶点与中心连线到凹进去的顶点与中心连线的偏移角度
    //恰好是 PI/5
    final double arc = Math.PI/5;
    //五角星凹进去的5个顶点对应的外接圆半径 rad
    final double rad = Math.cos(2 * Math.PI/5)/Math.cos(Math.PI/5);
    Path path = new Path();
    //以五角星的中心点为坐标系原点,以1为半径,从最右方的顶点开始绘制
    path.moveTo(1,0);
    for(int idx = 0;idx < 5;idx++){
        path.lineTo(
            (float)(rad * Math.cos((1 + 2 * idx) * arc)),
            (float)(rad * Math.sin((1 + 2 * idx) * arc)));
```

```
                    path.lineTo(
                        (float)(Math.cos(2 * (idx + 1) * arc)),
                        (float)(Math.sin(2 * (idx + 1) * arc)));
                }
                path.close();
                //创建转换矩阵
                Matrix matrix = new Matrix();
                //矩阵旋转角度:rotate
                matrix.postRotate(rotate);
                //缩放比例:radius
                matrix.postScale(radius, radius);
                //矩阵位移:水平方向 centerPointF.x,垂直方向 centerPointF.y
                matrix.postTranslate(centerPointF.x, centerPointF.y);
                //将矩阵转换效果运用在 path 中
                path.transform(matrix);
                return path;
            }
        }
```

(4) 定义 Activity

定义好使用的视图后,即可在 Activity 中显示。

参考代码如下:

```
        public class MainActivity extends Activity {
            FlagView fv;
            @Override
            protected void onCreate(Bundle savedInstanceState) {
                super.onCreate(savedInstanceState);
                fv = new FlagView(this);
                setContentView(fv);
            }
        }
```

7.1.2　常用的绘图类

1. Paint 的定义与使用

画笔 Paint 主要用来描述绘制图形的颜色和风格,如线宽、颜色、透明度和填充效果等。此类定义如下:

```
        java.lang.Object
          ↳ android.graphics.Paint
```

使用此类时,首先需要创建该类的对象,这可以通过该类提供的构造方法来实现。通常情况下,只需要使用无参数的构造方法来创建一个使用默认设置的 Paint 对象即可。

参考代码如下:

```
PaintbgPaint = new Paint();
```

创建 Paint 类对象后,可以通过其他方法对画笔的默认设置进行修改。常用方法如表 7-1 所示。

表 7-1 Paint 常用的方法说明

序号	方法名称	说明
1	public intgetColor()	返回画笔的颜色
2	public voidsetARGB(int a,int r,int g,int b)	设置画笔颜色,各参数值为 0~255 之间的整数,分别用来表示透明度、红色、绿色和蓝色值
4	public voidsetAlpha(int a)	设置透明度
5	public voidsetAntiAlias(boolean aa)	设置抗锯齿功能,如果使用会使绘图速度变慢
6	public voidsetColor(int color)	设置颜色,参数可以通过 Color 类提供的颜色常量指定,也可以通过 Color.rgb(int red,int green,int blue) 方法指定
7	public voidsetDither(boolean dither)	指定是否使用图像抖动处理,如果使用会使图像颜色更加平滑和饱满,使图像更加清晰
8	publicPathEffect setPathEffect(PathEffect effect)	设置绘制路径时的效果
9	publicShader setShader(Shader shader)	设置渐变,可以使用 LinearGradient、RadialGradient 和 SweepGradient
10	public voidsetShadowLayer(float radius,float dx,float dy,int color)	设置阴影,参数为阴影的角度 radius,阴影在 x 轴和 y 轴上的距离,阴影的颜色 color
11	public voidsetStrokeWidth(float width)	设置笔触宽度
12	public voidsetStyle(Paint.Style style)	设置填充风格
13	public voidsetTextAlign(Paint.Align align)	设置绘制文本时文字的对齐方式,参数为 Align.CENTER 或其他
14	public voidsetTextSize(float textSize)	设置绘制文本时文字的大小

2. Canvas 的定义与使用

画布 Canvas 主要用来绘制各种图形。此类定义如下:

```
java.lang.Object
  ↳ android.graphics.Canvas
```

一般地,在 Android 绘图时,需要先创建一个继承自 View 类的视图,并在该类中重写其 onDraw(Canvas c) 方法,然后在显示绘图的 Activity 中添加该视图。

创建好的视图也可以直接运用在布局文件中,需要注意的是,此时必须为自定义的 View 类实现如下构造方法:

```
public 类名(Context context,AttributeSet attributeSet){
    super(context,attributeSet);
}
```

例如,对于"绘制五星红旗"中 FlagView 类,必须实现以下构造方法:

```
publicFlagView(Context context,AttributeSet attributeSet){
    super(context,attributeSet);
}
```

然后，使用自定义 View 的完整路径及名称即可在布局文件 activity_main.xml 中运用此视图，例如，"绘制五星红旗"中 FlagView 类的路径（即包名）是"com.book.testdrawsim"，即可使用以下参考代码实现：

```
<?xml version = "1.0" encoding = "utf-8"?>
<LinearLayout xmlns:android = "http://schemas.android.com/apk/res/android"
    android:layout_width = "match_parent"
    android:layout_height = "match_parent"
    android:orientation = "vertical" >
<com.book.testdrawsim.FlagView
    android:layout_width = "wrap_content"
    android:layout_height = "wrap_content"
    />
</LinearLayout>
```

最后，在 MainActivity 中仍旧使用 setContentView(int layoutID) 的方法使用此布局。参考代码如下：

```
public classMainActivity extends Activity {
    @Override
    protected void onCreate(BundlesavedInstanceState) {
        super.onCreate(savedInstanceState);
        setContentView(R.layout.activity_main);
    }
}
```

7.1.3 绘制简单图形的基本方法

1. 绘制几何图形

常见的几何图形包括点、线、弧、圆形、矩形等。在 Android 中，Canvas 类提供了丰富的绘制方法，使用得当即可绘制各种几何图形。

Canvas 类中常用的绘制几何图形的方法如表 7-2 所示：

表 7-2　Canvas 类绘制几何图形的常用方法说明

序号	方法名称	说明
1	public void drawArc(RectF oval,float startAngle,float sweepAngle,boolean useCenter,Paint paint)	绘制弧，参数分别是矩形对象、弧的开始角度、对应的角度范围、是否显示中心角和画笔
2	public void drawCircle(float cx,float cy,float radius,Paint paint)	绘制圆，参数分别是圆心横、纵坐标，半径和画笔
3	public void drawLine(float startX,float startY,float stopX,float stopY,Paint paint)	绘制一每次线，参数分别是起点的横、纵坐标，终点的横、纵坐标和画笔

(续)

序号	方法名称	说 明
4	public void drawLines(float[ ] pts,Paint paint)	绘制多条线,参数分别是各条线的起点、终点坐标和画笔
5	public void drawPoint(float x,float y,Paint paint)	绘制一个点,参数分别是点的横、纵坐标和画笔
6	public void drawPoints(float[ ] pts,Paint paint)	绘制多个点,参数分别是各点的横、纵坐标和画笔
7	public void drawOval(RectF oval,Paint paint)	绘制椭圆,参数分别是矩形对象和画笔
8	public void drawRect(float left,float top,float right,float bottom,Paint paint)	绘制矩形,参数分别是矩形左上角顶点的横、纵坐标,右下角顶点的横、纵坐标和画笔
9	public void drawRoundRect(RectF rect,float rx,float ry,Paint paint)	绘制圆角矩形,参数分别是矩形对象、水平方向圆角与垂直方向圆角所在圆的半径和画笔

2. 绘制路径

除了上述规则的几何图形之外,在开发 Android 应用时,可能需要绘制一些不规则图形,此时可以运用 Canvas 类提供的绘制路径的方法完成。

绘制路径的基本步骤如下。

① 使用 Path 类创建路径对象。Path 类的定义如下:

```
java.lang.Object
  ↳ android.graphics.Path
```

创建 Path 类的对象时,可以通过该类提供的构造方法来实现。通常情况下,只需要使用无参数的构造方法来创建一个使用默认设置的 Path 对象即可。参考代码如下:

```
private PathmBigStarPath = new Path();
```

② 使用 Path 类的常用方法完成路径设置。Path 类的常用绘图方法如表 7-3 所示。

表 7-3 Canvas 类绘制路径的常用方法说明

序号	方法名称	说 明
1	public void addArc(RectF oval,float startAngle,float sweepAngle)	添加弧形路径,参数分别是矩形对象、开始角度和角度范围
2	public void addCircle(float x,float y,float radius,Path.Direction dir)	添加圆形路径,参数分别是圆心坐标、半径和路径绘制方向。Path.Direction.CCW 表示逆时针方向,Path.Direction.CW 表示顺时针方向
3	public void addOval(RectF oval,Path.Direction dir)	添加椭圆形路径,参数分别是矩形对象和绘制方向
4	public void addRect(float left,float top,float right,float bottom,Path.Direction dir)	添加矩形路径,参数分别是矩形左上角、右下角顶点的坐标和绘制方向
5	public void addRect(RectF rect,Path.Direction dir)	添加矩形路径,参数分别是矩阵对象和绘制方向
6	public void addRoundRect(RectF rect,float rx,float ry,Path.Direction dir)	添加圆角矩形路径,参数分别是矩形对象、水平方向圆角与垂直方向圆角所在圆的半径和绘制方向

(续)

序号	方法名称	说明
7	public void close()	闭合路径
8	public void lineTo(float x,float y)	在moveTo()起始点与该方法指定的结束点之间画一条线，起始点默认位置为(0,0)，参数分别是点的横、纵坐标
9	public void moveTo(float x,float y)	设置起始点，参数为点的横、纵坐标
10	public void transform(Matrix matrix)	将路径转置，参数为转置矩阵，参数是矩阵对象

③ 使用 Canvas 类的 drawPath(Path path, Paint paint) 方法完成路径绘制。参考代码如下：

```
canvas.drawPath(mBigStarPath,mStarPaint);
```

3. 绘制文本

虽然 Android 中提供了显示文本的组件，但在开发一些角色类游戏时，会包含很多文字，使用这些组件不太合适，此时可以通过 Canvas 类提供的绘制文本的方法来实现。绘制文本的基本方法如表 7-4 所示：

表 7-4 Canvas 类绘制文本的方法

序号	方法名称	说明
1	public void drawText(String text,float x,float y,Paint paint)	绘制文本，参数分别是字符串、文本的绘制起点坐标和画笔
2	public void drawText(String text,int start,int end,float x,float y,Paint paint)	绘制文本，参数分别是字符串、需要绘制的第1个字符位置、需要绘制的最后一个字符位置、文本的绘制起点坐标和画笔
3	public void drawTextOnPath(String text,Path path,float hOffset,float vOffset,Paint paint)	沿着路径绘制文本，参数分别是字符串、路径、文本的绘制起点位置相对于路径起点的偏移、文本与路径之间的距离和画笔

7.2 实例2：放大镜看SD卡中的图

7.2.1 功能要求与操作步骤

1. 功能要求

完成如图 7-2 所示的应用程序。要求能够：
① 运行应用程序，首先显示如图 7-2a 所示的放大镜。
② 单击"从图库中选择"按钮，可以显示图库中的图片，效果如图 7-2b 所示。
③ 当用户选择了图库中的某张图后，即可用放大镜查看放大效果，如图 7-2c 所示。

2. 操作步骤

（1）创建项目

参考如下输入信息创建 Android 项目：

图7-2 "放大镜看SD卡中的图"运行效果

```
Application Name: 07_TestMagnifyPhoto
Project Name: 07_TestMagnifyPhoto
Package Name: com.book.testmagnifyphoto
Activity Name: MainActivity
Layout Name: activity_main
```

（2）自定义视图 MagnifyView

从图7-2各图可看出，应用程序上方需要自定义视图，用来显示放大镜。另外，当用户从图库中选择好图片并显示在此区域后，当移动放大镜时，图片的局部将被放大。这些功能均在 MagnifyView 中定义。

参考代码如下：

```java
public class MagnifyView extends View {
    //从图库中获取到的图像
    private Bitmap bitmap;
    //放大后的圆形遮罩
    private ShapeDrawable drawable;
    //放大镜的半径,随着分辨率的不同会发生改变
    private final int RADIUS = 38;
    //放大倍数
    private final int FACTOR = 2;
    private Matrix matrix = new Matrix();
    //放大镜位图
    private Bitmap bitmap_magnifier;
    //放大镜的左边距
    private int m_left = 0;
    //放大镜的顶边距
    private int m_top = 0;
    //构造函数
    public MagnifyView(Context context, AttributeSet as) {
        super(context, as);
        //获取放大镜图像
```

```java
            bitmap_magnifier = BitmapFactory.decodeResource(
                    getResources(),
                    R.drawable.magnifier);
    }
    @Override
    protected void onDraw(Canvas canvas){
        super.onDraw(canvas);
        try{
            //绘制居中的背景图像
            canvas.drawBitmap(bitmap,
                    (getWidth() - bitmap.getWidth())/2,
                    (getHeight() - bitmap.getHeight())/2,
                    null);
            //绘制放大后的图像
            drawable.draw(canvas);
        }catch(Exception e){
            e.getCause();
        }
        //绘制放大镜
        canvas.drawBitmap(bitmap_magnifier,m_left,m_top,null);
    }
    @Override
    public boolean onTouchEvent(MotionEvent event){
        //获取当前触摸点的横坐标
        final int x = (int)event.getX();
        //获取当前触摸点的纵坐标
        final int y = (int)event.getY();
        try{
            //平移到绘制 shader 的起始位置
            matrix.setTranslate(RADIUS - x * FACTOR,RADIUS - y * FACTOR);
            drawable.getPaint().getShader().setLocalMatrix(matrix);
            //设置圆的外切矩形,以此触发 drawable 的绘制
            drawable.setBounds(x - RADIUS,y - RADIUS,x + RADIUS,y + RADIUS);
        }catch(Exception e){
            e.getCause();
        }
        //计算放大镜的左边距
        m_left = x - bitmap_magnifier.getWidth()/2;
        //计算放大镜的右边距
        m_top = y - bitmap_magnifier.getHeight()/2;
        invalidate();//重绘画布
        return true;
    }
    //获取要显示的源图,并放大图
    public void setBitmap(String picturePath){
        //取得源图
        bitmap = BitmapFactory.decodeFile(picturePath);
        //根据画布宽、高缩放源图
```

```
        bitmap = bitmap.createScaledBitmap(bitmap,getWidth(),getHeight(),true);
        //以 FACTOR 为倍数放大图
        magnifyBitmap(bitmap,FACTOR);
    }
    //生成放大效果图
    private void magnifyBitmap(Bitmap bitmap2,int factor){
        //TODO Auto-generated method stub
        //放大后的图
        BitmapShader shader = new BitmapShader(
                Bitmap.createScaledBitmap(
                        bitmap,
                        bitmap.getWidth() * factor,
                        bitmap.getHeight() * factor,
                        true),
                TileMode.CLAMP,
                TileMode.CLAMP);
        //圆形遮罩
        drawable = new ShapeDrawable(new OvalShape());
        //将放大后的图切除多余部分后放到 drawable 中
        drawable.getPaint().setShader(shader);
        //设置圆的外切矩形,以此触发 drawable 的绘制
        drawable.setBounds(0,0,RADIUS * 2,RADIUS * 2);
        //此处计算放大镜的边距可以使得遮罩和放大镜初始位置相同
        m_left = RADIUS - bitmap_magnifier.getWidth()/2;
        m_top = RADIUS - bitmap_magnifier.getHeight()/2;
    }
}
```

(3) 定义布局

自定义的视图 MagnifyView 和按钮构成了应用程序的主界面布局。这两个组件的属性如表 7-5 所示。

表 7-5 "放大镜看 SD 卡中的图"组件属性列表

序号	组件类型	属性名称	属性值	说明
1	LinearLayout	android:layout_width	match_parent	宽度匹配父容器
		android:layout_height	match_parent	高度匹配父容器
		android:orientation	vertical	垂直方向
2	com.book.testmagnifyphoto.MagnifyView	android:id	@+id/img_chosen	设置组件 id
		android:layout_width	match_parent	宽度匹配父容器
		android:layout_height	0dp	设置组件高度为 0,与权重配合使用
		android:layout_weight	9	占布局的 90%
3	Button	android:id	@+id/btn_choose	设置组件 id
		android:layout_width	match_parent	宽度匹配父容器

(续)

序号	组件类型	属性名称	属性值	说明
3	Button	android:layout_height	0dp	设置组件高度为0,与权重配合使用
		android:text	从图库中选择	设置显示文本
		android:layout_weight	1	占布局的10%
		android:onClick	clickChoose	设置单击事件名称

参考代码如下:

```xml
<?xml version = "1.0" encoding = "utf-8"?>
<LinearLayout xmlns:android = "http://schemas.android.com/apk/res/android"
    android:layout_width = "fill_parent"
    android:layout_height = "fill_parent"
    android:orientation = "vertical" >
<com.book.testmagnifyphoto.MagnifyView
    android:id = "@+id/img_chosen"
    android:layout_width = "match_parent"
    android:layout_height = "0dp"
    android:layout_weight = "9" />
<Button
    android:id = "@+id/btn_choose"
    android:layout_width = "match_parent"
    android:layout_height = "0dp"
    android:text = "从图库中选择"
    android:layout_weight = "2"
    android:onClick = "clickChoose" />
</LinearLayout>
```

(4) 在MainActivity中实现应用功能

在单击"从图库中选择"按钮时,需要将界面切换到图库界面,然后由用户选择一个图片,返回后将选择的图片显示在应用程序主界面中。

参考代码如下:

```java
public class MainActivity extends Activity {
    //定义请求码
    private static final int ChooseImage = 1;
    //声明 MagnifyView 对象
    MagnifyView imgChosen;
    @Override
    protected void onCreate(Bundle savedInstanceState) {
        super.onCreate(savedInstanceState);
        this.setContentView(R.layout.activity_main);
        imgChosen = (MagnifyView)findViewById(R.id.img_chosen);
    }
    public void clickChoose(View v) {
        //启动图库activity,并带回选择的图片
```

```
        Intent i = new Intent(
                Intent.ACTION_PICK,
                android.provider.MediaStore.Images.Media.EXTERNAL_CONTENT_URI);
        startActivityForResult(i,ChooseImage);
    }
    @Override
    protected void onActivityResult(int requestCode,int resultCode,Intent data){
        super.onActivityResult(requestCode,resultCode,data);
        //如果正确选择了图片
        if(requestCode == ChooseImage
                && resultCode == RESULT_OK
                &&data != null){
            //获取选择的图片
            Uri selectedImage = data.getData();
            String[] filePathColumn = { MediaStore.Images.Media.DATA };
            Cursor cursor = getContentResolver().query(
                    selectedImage,
                    filePathColumn,
                    null,
                    null,
                    null);
            cursor.moveToFirst();
            int columnIndex = cursor.getColumnIndex(filePathColumn[0]);
            String picturePath = cursor.getString(columnIndex);
            cursor.close();
            if(picturePath != null){
                imgChosen.setBitmap(picturePath);
            }
        }
    }
}
```

7.2.2 访问图库中的图像

1. 访问系统图库

在定义 Intent 时,使用 Intent(String action,Uri uri) 构造函数可以为 Intent 设置 action 和 Uri。将 Uri 设置为 "android.provider.MediaStore.Images.Media.EXTERNAL_CONTENT_URI" 时,即可访问系统图库。为了允许用户选择图库中的图片,需要将 action 设置为 "Intent.ACTION_PICK"。在定义好 Intent 后,再通过 startActivityForResult(Intent intent,int requestCode) 方法访问图库。

"启动图库 activity,并带回选择的图片" 的参考代码如下:

```
Intent i = new Intent(
    Intent.ACTION_PICK,
    android.provider.MediaStore.Images.Media.EXTERNAL_CONTENT_URI);
startActivityForResult(i,ChooseImage);
```

2. 使用 Bitmap 类缩放图片

位图 Bitmap 是 Android 系统中图像处理的一个重要的类。使用该类，不仅可以获取图像文件信息，进行图像剪切、旋转、缩放等操作，还可以指定格式保存图像文件。该类的定义如下：

```
java.lang.Object
 ↳ android.graphics.Bitmap
```

Bitmap 常用的方法如表 7-6 所示。

表 7-6　Bitmap 常用的方法说明

序号	方法名称	说　　明
1	public static Bitmap createBitmap(int width, int height, Bitmap.Config config)	用于创建一个指定宽度和高度的新的 Bitmap 对象
2	public static Bitmap createBitmap(Bitmap source, int x, int y, int width, int height, Matrix m, boolean filter)	用于从源位图的指定坐标点开始，"挖取"指定宽度和高度的一块图像来创建新的 Bitmap 对象，并按 Matrix 指定规则进行变换
3	public static Bitmap createBitmap(Bitmap source, int x, int y, int width, int height)	用于从源位图的指定坐标点开始，"挖取"指定宽度和高度的一块图像来创建新的 Bitmap 对象
4	public static Bitmap createBitmap(int[] colors, int width, int height, Bitmap.Config config)	使用颜色数组创建一个指定宽度和高度的新的 Bitmap 对象，其中数组元素的个数为 width * height
5	public static Bitmap createBitmap(Bitmap src)	使用源位图创建一个新的 Bitmap 对象
6	public static Bitmap createScaledBitmap(Bitmap src, int dstWidth, int dstHeight, boolean filter)	用于将源位图缩放为指定宽度和高度的新的 Bitmap 对象

例如，在"放大镜看 SD 卡中的图"实例中，为了使 bitmap 图像在画布中平铺显示，需要按画布大小（宽为 getWidth()，高为 getHeight()）缩放源图，使用的代码如下：

```
bitmap = bitmap.createScaledBitmap(bitmap, getWidth(), getHeight(), true);
```

再如，为了将 bitmap 图像放大 factor 倍，使用了如下代码：

```
bitmap.createScaledBitmap(bitmap,
              bitmap.getWidth() * factor,
              bitmap.getHeight() * factor,
              true)
```

3. 使用 BitmapFactory 类解析、创建 Bitmap 图形对象

BitmapFactory 类是一个工具类，用于从不同的数据源来解析、创建 Bitmap 对象。该类的定义如下：

```
java.lang.Object
 ↳ android.graphics.BitmapFactory
```

BitmapFactory 常用的方法如表 7-7 所示。

表 7-7　BitmapFactory 常用的方法说明

序号	方法名称	说　　明
1	public static Bitmap decodeFile(String pathName)	从给定的路径所指定的文件中解析、创建 Bitmap 对象

(续)

序号	方法名称	说明
2	public static Bitmap decodeFileDescriptor (FileDescriptor fd)	从 FileDescriptor 对应的文件中解析、创建 Bitmap 对象
3	public static Bitmap decodeResource (Resources res , int id)	根据给定的资源 id 解析、创建 Bitmap 对象
4	public static Bitmap decodeStream(InputStream is)	从指定的输入流中解析、创建 Bitmap 对象

例如，在"放大镜看 SD 卡中的图"实例中，首先从图库中取出图片路径 picturePath，然后根据此路径，解析、创建 Bitmap 对象，使用的代码如下：

```
bitmap = BitmapFactory. decodeFile( picturePath ) ;
```

再如，放大镜的图片存放在 res/drawable/magnifier. png，在自定义的 MagnifyView 中绘制放大镜时，使用了如下代码：

```
bitmap_magnifier = BitmapFactory. decodeResource(
        getResources( ) ,
        R. drawable. magnifier ) ;
```

7.3 实例 3：扣篮瞬间

7.3.1 功能要求与操作步骤

1. 功能要求

完成如图 7-3 所示的应用程序。要求能够：

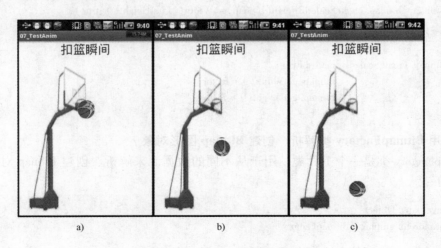

图 7-3 "扣篮瞬间"运行效果图

① 在应用程序界面中，上方居中显示文字"扣篮瞬间"，左边显示一个篮球架。
② 篮球从篮框处旋转着加速落地，并循环播放此动画过程。

2. 操作步骤

（1）创建项目

参考如下输入信息创建 Android 项目：

> Application Name：07_TestAnim
> Project Name：07_TestAnim
> Package Name：com. book. testanim
> Activity Name：MainActivity
> Layout Name：activity_main

（2）定义布局

本应用布局可以用嵌套线性布局实现，在外层线性布局垂直排列文本显示组件和内层线性布局。在内层线性布局中放置水平方向的两个图片框。所有组件的相关属性可参考表7-8设置。

表7-8 "扣篮瞬间"组件属性列表

序号	组件类型	属性名称	属性值	说明
1	LinearLayout（外层）	android：layout_width	match_parent	宽度匹配父容器
		android：layout_height	match_parent	高度匹配父容器
		android：orientation	vertical	垂直方向
2	TextView	android：layout_width	wrap_content	宽度适合内容
		android：layout_height	wrap_content	高度适合内容
		android：layout_gravity	center	居中放置文本框
		android：text	扣篮瞬间	设置显示文本
		android：textSize	30sp	设置文本大小
3	LinearLayout（内层）	android：layout_width	match_parent	宽度匹配父容器
		android：layout_height	match_parent	高度匹配父容器
4	ImageView（篮球架）	android：layout_width	wrap_content	宽度适合内容
		android：layout_height	400dp	设置高度
		android：scaleType	fitXY	将图片按图片框大小缩放
		android：src	@drawable/basketball_net	设置显示图片
5	ImageView（篮球）	android：id	@+id/img_earth	设置组件id
		android：layout_width	50dp	设置组件宽度
		android：layout_height	50dp	设置组件高度
		android：src	@drawable/lanqiu	设置显示图片

参考代码如下：

> \< LinearLayout xmlns：android = "http：//schemas. android. com/apk/res/android"
>   android：layout_width = " match_parent"
>   android：layout_height = " match_parent"
>   android：orientation = " vertical" \>
> \< TextView

```
            android:layout_width = "wrap_content"
            android:layout_height = "wrap_content"
            android:layout_gravity = "center"
            android:text = "扣篮瞬间"
            android:textSize = "30sp"/>
    <LinearLayout
            android:layout_width = "match_parent"
            android:layout_height = "match_parent" >
        <ImageView
                android:layout_width = "wrap_content"
                android:layout_height = "400dp"
                android:scaleType = "fitXY"
                android:src = "@drawable/basketball_net"/>
        <ImageView
                android:id = "@+id/img_ball"
                android:layout_width = "50dp"
                android:layout_height = "50dp"
                android:src = "@drawable/lanqiu"/>
    </LinearLayout>
</LinearLayout>
```

（3）制作篮球自身的旋转动画 self.xml

篮球在被扣下的时候处于旋转的状态，假设篮球是沿球心（android:pivotX = 50%, android:pivotY = 50%）顺时针旋转（从0°旋转到360°），旋转一周需要300ms的时间，动画无限循环播放。此动画设置文件存放在 res/anim/self.xml 中。

参考代码如下：

```
<?xml version = "1.0" encoding = "utf-8"?>
<rotate
    xmlns:android = "http://schemas.android.com/apk/res/android"
    android:interpolator = "@android:anim/accelerate_interpolator"
    android:fromDegrees = "0"
    android:toDegrees = "360"
    android:repeatCount = "infinite"
    android:pivotX = "50%"
    android:pivotY = "50%"
    android:duration = "300" >
</rotate>
```

（4）制作篮球从篮框垂直落下的动画 down.xml

篮球从篮框垂直落下时，使用的是平移动画。指定篮球落下的起点和终点，以及循环模式和动画时间即可。此动画设置文件存放在 res/anim/down.xml 中。

参考代码如下：

```
<?xml version = "1.0" encoding = "utf-8"?>
<translate
    xmlns:android = "http://schemas.android.com/apk/res/android"
```

```xml
android:fromXDelta = " -50"
android:toXDelta = " -50"
android:fromYDelta = "70"
android:toYDelta = "300"
android:repeatCount = "infinite"
android:repeatMode = "restart"
android:interpolator = "@android:anim/accelerate_interpolator"
android:duration = "1000" >
</translate>
```

（5）在 MainActivity 中播放动画

将动画设置好后，即可在 Activity 类中为指定对象添加动画，并播放。本例中，需要为篮球对象（imgBall）添加两个动画。为了便于管理，可以在类中定义一个动画集合 AnimationSet 对象，将篮球对象需要播放的所有动画全都添加到该对象中。

参考代码如下：

```java
public class MainActivity extends Activity {
    //声明图片组件
    ImageView imgBall;
    @Override
    protected void onCreate(Bundle savedInstanceState) {
        super.onCreate(savedInstanceState);
        setContentView(R.layout.activity_main);
        //定义图片组件对象
        imgBall = (ImageView)findViewById(R.id.img_ball);
        //加载旋转动画
        Animation rotate = AnimationUtils.loadAnimation(this, R.anim.self);
        //加载平移动画
        Animation transferm = AnimationUtils.loadAnimation(this, R.anim.down);
        //定义动画集合
        AnimationSet as = new AnimationSet(true);
        //将旋转动画和平移动画都添加到动画集合中
        as.addAnimation(rotate);
        as.addAnimation(transferm);
        //图片对象播放动画集合中的所有动画
        imgBall.startAnimation(as);
    }
}
```

7.3.2 Android 动画技术简介

1. 旋转动画

旋转动画是通过为动画指定开始时的旋转角度、结束时的旋转角度以及持续时间来创建动画。在旋转时，还可以通过指定轴心点坐标来改变旋转的中心。可以在 XML 文件中定义旋转动画资源文件，基本的语法格式可参考如下代码：

```xml
<?xml version="1.0" encoding="utf-8"?>
<rotate
    xmlns:android="http://schemas.android.com/apk/res/android"
    android:interpolator="@android:anim/accelerate_interpolator"
    android:fromDegrees="0"
    android:toDegrees="360"
    android:repeatCount="infinite"
    android:pivotX="50%"
    android:pivotY="50%"
    android:duration="300" >
</rotate>
```

在上述代码中,使用到的各种动画属性及其说明如表7-9所示。

表7-9 旋转动画常用属性及说明

序号	属性名称	说明
1	android:interpolator	设置动画的加速度,其可选值见表7-10
2	android:fromDegrees	指定动画开始时的旋转角度
3	android:toDegrees	指定动画结束时的旋转角度
4	android:repeatCount	设置动画的重复次数,可以是数值,也可以是infinite(即无限循环)
5	android:pivotX	指定轴心点的X坐标
6	android:pivotY	指定轴心点的Y坐标
7	android:duration	指定动画持续的时间,单位是毫秒

设置android:interpolator属性可以使得动画以匀速、加速、减速或抛物线速度播放,可选值如表7-10所示。

表7-10 android:interpolator可选值列表

序号	值	说明
1	@android:anim/accelerate_interpolator	加速播放动画
2	@android:anim/accelerate_decelerate_interpolator	先加速后减速再加速播放动画
3	@android:anim/decelerate_interpolator	减速播放动画
4	@android:anim/anticipate_interpolator	在动画开始的地方先向后退一小步,再开始动画
5	@android:anim/overshoot_interpolator	当动画结束时超出一小步再回到动画结束的地方
6	@android:anim/anticipate_overshoot_interpolator	在动画开始的地方先向后退一小步,再开始动画,当动画结束时超出一小步再回到动画结束的地方
7	@android:anim/bounce_interpolator	动画结束的地方采用弹球效果
8	@android:anim/linear_interpolator	匀速播放动画
9	@android:anim/cycle_interpolator	动画循环播放特定的次数,变化速度按正弦曲线改变

2. 平移动画

平移动画是通过为动画指定起止位置以及持续时间来创建动画,与旋转动画一样,也可以在XML文件中定义,基本的语法格式可参考如下代码:

```xml
<?xml version = "1.0" encoding = "utf-8"?>
<translate
    xmlns:android = "http://schemas.android.com/apk/res/android"
    android:fromXDelta = "-50"
    android:toXDelta = "-50"
    android:fromYDelta = "70"
    android:toYDelta = "300"
    android:repeatCount = "infinite"
    android:repeatMode = "restart"
    android:interpolator = "@android:anim/accelerate_interpolator"
    android:duration = "1000" >
</translate>
```

在上述代码中，使用到了 android:repeatCount、android:duration 属性，其用法与旋转动画中的相同，其他属性及其说明如表 7-11 所示。

表 7-11 平移动画常用属性说明

序号	属性名称	说明
1	android:fromXDelta	动画开始时水平方向上的位置
2	android:toXDelta	动画结束时水平方向上的位置
3	android:fromYDelta	动画开始时垂直方向上的位置
4	android:toYDelta	动画结束时垂直方向上的位置
5	android:repeatMode	动画的重复方式，可以是 reverse（反向）或 restart（重新开始）

3. 缩放动画

缩放动画是通过为动画指定起止时的缩放系数以及持续时间来创建动画。在缩放时，还可以通过指定轴心点坐标来改变缩放的中心。同样，也可以在 XML 文件中定义缩放动画资源文件，基本的语法格式可参考如下代码：

```xml
<?xml version = "1.0" encoding = "utf-8"?>
<scale
    xmlns:android = "http://schemas.android.com/apk/res/android"
    android:fromXScale = "1.0"
    android:toXScale = "3.0"
    android:fromYScale = "1.0"
    android:toYScale = "3.0"
    android:pivotX = "50%"
    android:pivotY = "50%"
    android:duration = "2000"
    android:repeatMode = "reverse" >
</scale>
```

在上述代码中，android:repeatCount、android:duration、android:pivotX、android:pivotY、android:repeatMode 等属性的用法均与其他动画中的相同，其余常用属性及其说明如表 7-12 所示。

表7-12 缩放动画常用属性说明

序号	属性名称	说明
1	android:fromXScale	动画开始时水平方向上的缩放系数
2	android:toXScale	动画结束时水平方向上的缩放系数
3	android:fromYScale	动画开始时垂直方向上的缩放系数
4	android:toYScale	动画结束时垂直方向上的缩放系数

4. 透明度渐变动画

透明度渐变动画就是通过 View 组件透明度的变化来实现其渐隐渐现的效果。可以使用设定动画开始和结束时的透明度及持续时间来创建动画。与其他动画一样，该动画也可以在 XML 文件中定义，基本的语法格式可参考如下代码：

```
<?xml version = "1.0" encoding = "utf-8"?>
<alpha
    xmlns:android = "http://schemas.android.com/apk/res/android"
    android:fromAlpha = "0"
    android:toAlpha = "1"
    android:duration = "1000"/>
```

在上述代码中，使用到了两个特殊的属性 android:fromAlpha 和 android:toAlpha，使用说明如表 7-13 所示。

表7-13 旋转动画常用属性及说明

序号	属性名称	说明
1	android:fromAlpha	指定动画开始时的透明度，值为0表示完全透明，值为1表示完全不透明
2	android:toAlpha	指定动画结束时的透明度，值为0表示完全透明，值为1表示完全不透明

7.4 实例4：悦视播放器

7.4.1 功能要求与操作步骤

1. 功能要求

完成如图 7-4 所示的应用程序。要求能够：

① 应用程序首界面如图 7-4a 所示。

② 单击首界面中的"播放音频"按钮，切换到显示 SD 卡中所有音频文件列表界面，如图 7-4b 所示。单击任意列表项即可播放该音频文件，此后界面下方的"Back""Pause"以及"Next"按钮可以使用。当单击系统的"回退"键时，音频结束播放。

③ 单击首界面中的"播放视频"按钮，切换到如图 7-4d 所示的界面。单击"播放"按钮，即可播放指定视频，在播放过程中可以随时"暂停"和"停止"播放，如图 7-4e 所示。当视频播放结束后，界面切换到首界面，并提示"视频播放完毕"，如图 7-4f 所示。

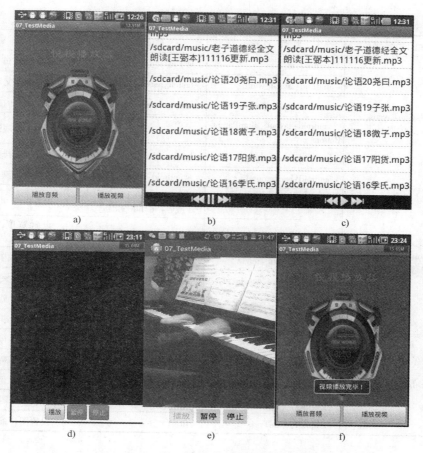

图 7-4 "悦视播放器"运行效果图

2. 操作步骤

（1）创建项目

参考如下输入信息创建 Android 项目：

> Application Name：07_TestMedia
> Project Name：07_TestMedia
> Package Name：com. book. testmedia
> Activity Name：MainActivity
> Layout Name：activity_main

（2）定义 activity_main.xml 布局

从图 7-4a 可看出，该布局文件用嵌套布局将图片和两个按钮组件合理放置即可。各个组件的属性可以参考表 7-14 设置。

表 7-14 "悦视播放器"首界面组件属性列表

序号	组件类型	属性名称	属性值	说　　明
1	RelativeLayout（外层）	android:layout_width	match_parent	宽度匹配父容器
		android:layout_height	match_parent	高度匹配父容器

233

(续)

序号	组件类型	属性名称	属性值	说明
2	LinearLayout（内层）	android:id	@+id/ll_function	设置组件 id
		android:layout_width	match_parent	宽度匹配父容器
		android:layout_height	wrap_content	高度适合内容
		android:layout_alignParentBottom	true	与父容器底部对齐
		android:orientation	horizontal	水平方向排列组件
		android:background	@android:color/darker_gray	设置背景颜色
3	Button（"播放音频"和"播放视频"按钮）	android:id	@+id/btn_audio（或@+id/btn_video）	设置组件 id
		android:layout_width	0dp	设置宽度，与权重配合使用
		android:layout_height	wrap_content	高度适合内容
		android:layout_weight	5	宽度各占父容器的 50%
		android:text	播放音频（或播放视频）	设置显示文本
		android:onClick	clickAudio（或 clickVideo）	设置单击事件名称
4	ImageView	android:layout_above	@+id/ll_function	设置组件 id
		android:layout_height	match_parent	设置组件高度
		android:layout_width	match_parent	设置组件宽度
		android:src	@drawable/player	设置显示图片
		android:scaleType	fitXY	设置图片缩放比例

参考代码如下：

```xml
<RelativeLayout xmlns:android = "http://schemas.android.com/apk/res/android"
    xmlns:tools = "http://schemas.android.com/tools"
    android:layout_width = "match_parent"
    android:layout_height = "match_parent" >
    <LinearLayout
        android:id = "@+id/ll_function"
        android:layout_width = "match_parent"
        android:layout_height = "wrap_content"
        android:layout_alignParentBottom = "true"
        android:orientation = "horizontal"
        android:background = "@android:color/darker_gray" >
        <Button
            android:id = "@+id/btn_audio"
            android:layout_width = "0dp"
            android:layout_height = "wrap_content"
            android:layout_weight = "5"
            android:text = "播放音频"
            android:onClick = "clickAudio" />
```

```xml
<Button
    android:id = "@+id/btn_video"
    android:layout_width = "0dp"
    android:layout_height = "wrap_content"
    android:text = "播放视频"
    android:layout_weight = "5"
    android:onClick = "clickVideo"/>
</LinearLayout>
<ImageView
    android:layout_width = "match_parent"
    android:layout_above = "@+id/ll_function"
    android:layout_height = "match_parent"
    android:src = "@drawable/player"
    android:scaleType = "fitXY"/>
</RelativeLayout>
```

（3）定义播放音频的 audio_layout.xml 布局

从图 7-4b 可看出，该布局与 activity_main.xml 的布局很相似。各个组件的属性可以参考表 7-15 设置。

表 7-15 "悦视播放器"播放音频的布局组件属性列表

序号	组件类型	属性名称	属性值	说明
1	LinearLayout（外层）	android:layout_width	match_parent	宽度匹配父容器
		android:layout_height	match_parent	高度匹配父容器
		android:orientation	vertical	垂直方向排列组件
2	ListView	android:id	@+id/list	设置组件id
		android:layout_width	fill_parent	宽度匹配父容器
		android:layout_height	0dp	高度为0，与权重配合使用
		android:layout_weight	9	高度占父容器90%
3	LinearLayout（内层）	android:layout_width	match_parent	宽度匹配父容器
		android:layout_height	0dp	高度为0，与权重配合使用
		android:layout_weight	1	高度占父容器的10%
		android:gravity	center	内部组件居中摆放
		android:background	@android:color/background_dark	设置背景颜色
4	Button（上一首、暂停和下一首）	android:id	@+id/pre（或@+id/pause、@+id/next）	设置组件id
		android:layout_width	wrap_content	宽度适合内容
		android:layout_height	wrap_content	高度适合内容
		android:enabled	false	按钮不可用

(续)

序号	组件类型	属性名称	属性值	说明
4	Button(上一首、暂停和下一首)	android:background	@android:drawable/ic_media_previous（或@android:drawable/ic_media_pause、@android:drawable/ic_media_next）	设置背景图片

参考代码如下：

```xml
<?xml version="1.0" encoding="utf-8"?>
<LinearLayout xmlns:android="http://schemas.android.com/apk/res/android"
    android:layout_width="fill_parent"
    android:layout_height="fill_parent"
    android:orientation="vertical"
    >
<ListView
    android:id="@+id/list"
    android:layout_width="fill_parent"
    android:layout_height="0dp"
    android:layout_weight="9"
    />
<LinearLayout
        android:layout_width="match_parent"
        android:layout_height="0dp"
        android:layout_weight="1"
        android:gravity="center"
        android:background="@android:color/background_dark" >
<Button
        android:id="@+id/pre"
        android:layout_width="wrap_content"
        android:layout_height="wrap_content"
        android:enabled="false"
        android:background="@android:drawable/ic_media_previous"/>
<Button
    android:id="@+id/pause"
    android:layout_width="wrap_content"
    android:layout_height="wrap_content"
    android:enabled="false"
    android:background="@android:drawable/ic_media_pause"/>
<Button
    android:id="@+id/next"
    android:layout_width="wrap_content"
    android:layout_height="wrap_content"
    android:enabled="false"
    android:background="@android:drawable/ic_media_next"/>
</LinearLayout>
</LinearLayout>
```

(4) 定义播放视频的 video_layout.xml 布局

从图 7-4d 可看出，该布局与 activity_main.xml 的布局也很相似。各个组件的属性列表不再详述。参考代码如下：

```xml
<?xml version = "1.0" encoding = "utf-8"?>
<LinearLayout xmlns:android = "http://schemas.android.com/apk/res/android"
    android:layout_width = "fill_parent"
    android:layout_height = "fill_parent"
    android:gravity = "center"
    android:orientation = "vertical" >
    <SurfaceView
        android:id = "@+id/surfaceView"
        android:keepScreenOn = "true"
        android:layout_width = "match_parent"
        android:layout_height = "0dp"
        android:layout_weight = "9" />
    <LinearLayout
        android:layout_width = "wrap_content"
        android:layout_height = "0dp"
        android:layout_weight = "1" >
        <Button
            android:id = "@+id/play"
            android:layout_width = "wrap_content"
            android:layout_height = "wrap_content"
            android:enabled = "false"
            android:text = "播放"
            android:onClick = "clickPlay" />
        <Button
            android:id = "@+id/pause"
            android:layout_width = "wrap_content"
            android:layout_height = "wrap_content"
            android:enabled = "false"
            android:text = "暂停"
            android:onClick = "clickPause" />
        <Button
            android:id = "@+id/stop"
            android:layout_width = "wrap_content"
            android:layout_height = "wrap_content"
            android:enabled = "false"
            android:text = "停止"
            android:onClick = "clickStop" />
    </LinearLayout>
</LinearLayout>
```

(5) 在 AudioActivity.java 中实现音频播放

"播放音频"模块有以下子功能需要实现：

① 使用 getAudioFiles(String url) 方法获取 SD 卡中的全部音频文件，并将这些文件路径存放到列表对象 audioList 中。

② 将 audioList 对象中的全部元素显示在列表组件 list 中。

③ 为列表组件 list 添加列表项单击事件，当单击时记录音频文件的当前位置，播放当前音频文件，并使"上一首""暂停"和"下一首"按钮可用。

④ 为 MediaPlayer 添加完成事件监听器，播放完当前曲目后播放下一首。

⑤ 为 Activity 不同的生命周期状态设置相应的播放状态。如当有电话打进时，音频应暂停播放，当接打完毕后再继续播放。

参考代码如下：

```java
public class AudioActivity extends Activity {
    //声明 MediaPlayer 对象
    private MediaPlayer mediaPlayer;
    //要播放的音频列表
    private List<String> audioList = new ArrayList<String>();
    //当前播放歌曲的索引
    private int currentItem = 0;
    //声明按钮对象
    private Button pause, pre, next;
    //合法的音频文件格式
    private static String[] audioFormatSet = new String[]{"mp3","wav","3gp"};
    @Override
    public void onCreate(Bundle savedInstanceState) {
        super.onCreate(savedInstanceState);
        setContentView(R.layout.audio_layout);
        //实例化一个 MediaPlayer 对象
        mediaPlayer = new MediaPlayer();
        //获取按钮
        pause = (Button)findViewById(R.id.pause);
        pre = (Button)findViewById(R.id.pre);
        next = (Button)findViewById(R.id.next);
        //使用 ListView 组件显示 SD 卡上的全部音频文件
        showAudioList();
        //为 MediaPlayer 添加完成事件监听器
        mediaPlayer.setOnCompletionListener(new OnCompletionListener(){
            @Override
            public void onCompletion(MediaPlayer mp){
                //播放下一首
                nextMusic();
            }
        });
        //为"暂停"按钮添加单击事件监听器
        pause.setOnClickListener(new OnClickListener(){
            @Override
            public void onClick(View v){
                if(mediaPlayer.isPlaying()){
                    //暂停音频的播放
                    mediaPlayer.pause();
                    pause.setBackgroundResource(android.R.drawable.ic_media_play);
```

```java
                } else {
                    //继续播放
                    mediaPlayer.start();
                    pause.setBackgroundResource(android.R.drawable.ic_media_pause);
                }
            }
        });
        //为"下一首"按钮添加单击事件监听器
        next.setOnClickListener(new OnClickListener() {
            @Override
            public void onClick(View v) {
                //播放下一首
                nextMusic();
            }
        });
        //为"上一首"按钮添加单击事件监听器
        pre.setOnClickListener(new OnClickListener() {
            //播放上一首
            @Override
            public void onClick(View v) {
                preMusic();
            }
        });
    }
    @Override
    protected void onPause() {
        //TODO Auto-generated method stub
        super.onPause();
        mediaPlayer.pause();
    }
    @Override
    protected void onResume() {
        //TODO Auto-generated method stub
        super.onResume();
        mediaPlayer.start();
    }
    @Override
    protected void onDestroy() {
        if(mediaPlayer.isPlaying()) {
            //停止音乐的播放
            mediaPlayer.stop();
        }
        //释放资源
        mediaPlayer.release();
        //结束当前活动
        this.finish();
        super.onDestroy();
    }
}
```

```java
//使用 ListView 组件显示 SD 卡上的全部音频文件
private void showAudioList(){
    //获取 SD 卡上的全部音频文件
    getAudioFiles("/sdcard/");
    //创建适配器
    ArrayAdapter<String> adapter = new ArrayAdapter<String>(this,
            android.R.layout.simple_list_item_1,audioList);
    //获取布局管理器中添加的 ListView 组件
    ListView listview = (ListView)findViewById(R.id.list);
    //将适配器与 ListView 关联
    listview.setAdapter(adapter);
    //当单击列表项时播放音乐
    listview.setOnItemClickListener(new OnItemClickListener(){
        @Override
        public void onItemClick(AdapterView<?> listView,View view,
                int position,long id){
            //将当前列表项的索引值赋值给 currentItem
            currentItem = position;
            //调用 playMusic()方法播放音乐
            playMusic(audioList.get(currentItem));
        }
    });
}
//取得合法文件
private void getAudioFiles(String url){
    //创建文件对象
    File files = new File(url);
    File[] file = files.listFiles();
    try{
        //遍历获取到的文件数组
        for(File f : file){
            //如果是目录,也就是文件夹
            if(f.isDirectory()){
                //递归调用
                getAudioFiles(f.getAbsolutePath());
            }else{
                //如果是音频文件
                if(isAudioFile(f.getPath())){
                    //将文件的路径添加到 list 集合中
                    audioList.add(f.getPath());
                }
            }
        }
    }catch(Exception e){
        e.printStackTrace();
    }
}
//判断是否为音频文件
```

```java
private static boolean isAudioFile(String path){
    //遍历合法格式数组
    for(String format : audioFormatSet){
        //判断是否为合法的音频文件
        if(path.contains(format)){
            return true;
        }
    }
    return false;
}
//播放音频
void playMusic(String path){
    try{
        //停止当前音频的播放
        if(mediaPlayer.isPlaying()){
            mediaPlayer.stop();
        }
        //重置 MediaPlayer
        mediaPlayer.reset();
        //指定要播放的音频文件
        mediaPlayer.setDataSource(path);
        //预加载音频文件
        mediaPlayer.prepare();
        //播放音频
        mediaPlayer.start();
        //设置各按钮可用
        pause.setEnabled(true);
        pre.setEnabled(true);
        next.setEnabled(true);
        //在播放时,暂停按钮的背景图片为 pause
        pause.setBackgroundResource(android.R.drawable.ic_media_pause);
    } catch(Exception e){
        e.printStackTrace();
    }
}
//下一首
void nextMusic(){
    //当对 currentItem 进行 +1 操作后,如果其值大于等于音频文件的总数
    if(++currentItem >= audioList.size()){
        currentItem = 0;
    }
    //调用 playMusic()方法播放音乐
    playMusic(audioList.get(currentItem));
}
//上一首
void preMusic(){
    //当对 currentItem 进行 -1 操作后,如果其值大于等于 0
    if(--currentItem >= 0){
```

```
            //如果currentItem的值大于等于音频文件的总数
            if(currentItem >= audioList.size()){
                currentItem = 0;
            }
        } else {
            //currentItem的值设置为音频文件总数-1
            currentItem = audioList.size()-1;
        }
        //调用playMusic()方法播放音乐
        playMusic(audioList.get(currentItem));
    }
}
```

(6) 在VideoActivity.java中实现视频播放

"播放视频"模块中需要注意的是为了使视频中的画面和声音同时播放，必须在播放视频前先将SurfaceView准备好。另外需要注意各个按钮之间的关系，例如只有当视频在播放时，"暂停"和"停止"按钮才可以使用；当视频在暂停时，"播放"按钮不可用，"停止"按钮可用，如图7-5所示。

图7-5 暂停视频播放的效果图

参考代码如下：

```
public class VideoActivity extends Activity implements Callback {
    //声明 MediaPlayer 对象
    private MediaPlayer mp;
    //声明 SurfaceView 对象和其他组件
    private SurfaceView sv;
    Button play,pause,stop;
    private SurfaceHolder surfaceHolder;
    @Override
    public void onCreate(Bundle savedInstanceState){
        super.onCreate(savedInstanceState);
        setContentView(R.layout.video_layout);
        //实例化 MediaPlayer 对象
```

```java
        mp = new MediaPlayer();
        //获取布局管理器中添加的 SurfaceView 组件
        sv = (SurfaceView)findViewById(R.id.surfaceView);
        //准备好 surfaceHolder
        prepareSurfaceHolder();
        //获取按钮
        play = (Button)findViewById(R.id.play);
        pause = (Button)findViewById(R.id.pause);
        stop = (Button)findViewById(R.id.stop);
        //为 MediaPlayer 对象添加完成事件监听器
        mp.setOnCompletionListener(new OnCompletionListener(){
            @Override
            public void onCompletion(MediaPlayer mp){
                Toast.makeText(VideoActivity.this,"视频播放完毕!",
                    Toast.LENGTH_SHORT).show();
                Intent i = new Intent(VideoActivity.this,MainActivity.class);
                startActivity(i);
                finish();
            }
        });
    }
    //当 Activity 处于 onResume 的状态时,重新准备 surfaceHolder
    @Override
    protected void onResume(){
        super.onResume();
        prepareSurfaceHolder();
    }
    //在当前 Activity 销毁时停止播放视频并释放相关资源
    @Override
    protected void onDestroy(){
        if(mp.isPlaying()){
            //停止播放视频
            mp.stop();
        }
        //释放资源
        mp.release();
        this.finish();
        super.onDestroy();
    }
    //准备 surfaceHolder
    private void prepareSurfaceHolder(){
        //TODO Auto-generated method stub
        surfaceHolder = sv.getHolder();
        surfaceHolder.setType(SurfaceHolder.SURFACE_TYPE_PUSH_BUFFERS);
        surfaceHolder.addCallback(this);
    }
    //在 surfaceView 创建好,surfaceHolder 准备好之后使"播放"按钮可用
    @Override
```

```java
public void surfaceCreated(SurfaceHolder holder){
    play.setEnabled(true);
}
public void clickPlay(View v){
    //重置 MediaPlayer 对象
    mp.reset();
    try{
        //设置要播放的视频路径
        mp.setDataSource("/sdcard/Download/2964886.mp4");
        //设置将视频画面输出到 SurfaceView
        mp.setDisplay(sv.getHolder());
        //预加载视频
        mp.prepare();
        //开始播放
        mp.start();
        pause.setText("暂停");
        //设置按钮可用状态
        pause.setEnabled(true);
        stop.setEnabled(true);
        play.setEnabled(false);
    }catch(Exception e){
        e.printStackTrace();
    }
}
public void clickPause(View v){
    if(mp.isPlaying()){
        //暂停视频的播放
        mp.pause();
        ((Button)v).setText("继续");
    }else{
        //继续视频的播放
        mp.start();
        ((Button)v).setText("暂停");
    }
}
public void clickStop(View v){
    if(mp.isPlaying()){
        //停止播放
        mp.stop();
        //设置按钮不可用
        pause.setEnabled(false);
        stop.setEnabled(false);
        play.setEnabled(true);
    }
}
@Override
public void surfaceChanged(SurfaceHolder holder,int format,int width,
        int height){
```

```
        }
        @Override
        public void surfaceDestroyed(SurfaceHolder holder){
        }
    }
```

7.4.2 音频的播放与录制

1. 使用 MediaPlayer 播放音频

MediaPlayer 类为控制音频和视频提供了播放、暂停、停止和重复播放等方法。该类的定义如下：

```
java.lang.Object
  ↳ android.media.MediaPlayer
```

MediaPlayer 常用的方法如表 7-16 所示。

表 7-16 MediaPlayer 常用的方法列表

序号	方法名称	说明
1	public MediaPlayer()	无参构造函数，用来创建 MediaPlayer 对象
2	public static MediaPlayer create(Context context, int resid)	用资源 id 对应的资源文件装载音频，返回新创建的 MediaPlayer 对象
3	public static MediaPlayer create(Context context, Uri uri)	根据指定的 uri 装载音频，返回新创建的 MediaPlayer 对象
4	public boolean isPlaying()	判断是否在播放中
5	public void pause()	暂停播放
6	public void prepare()	预加载文件
7	public void release()	释放音频资源
8	public void reset()	重新设置音频
9	public void seekTo(int msec)	找到指定时间的位置
10	public void setDataSource(String path)	设置音频文件
11	public void setDisplay(SurfaceHolder sh)	设置视频显示区
12	public void start()	开始播放
13	public void stop()	停止播放

在使用 MediaPlayer 播放音、视频时，首先要创建该类的对象，并装载音频文件。有两种方法实现。

① 使用 MediaPlayer 类的静态方法 create() 方法来创建对象，并装载音频文件。此时，在对象创建的同时，已经装载了要播放的音频。

例如，构造 Music 类，其中的 play(Context ctxt, int resId) 方法用来播放"res/raw"目录中的音频资源文件；pause(Context ctxt) 方法用来暂停播放音频；stop(Context ctxt) 方法用来停止音频播放，并释放音频资源。

参考代码如下：

```
public class Music {
    private static MediaPlayer mp = null;
    public static void play(Contextctxt, int resId) {
        stop(ctxt);
        mp = MediaPlayer.create(ctxt, resource);
        mp.setLooping(true);
        mp.start();
    }
    public static void pause(Contextctxt) {
        mp.pause();
    }
    public static void stop(Contextctxt) {
        //TODO Auto - generated method stub
        if(mp != null) {
            mp.stop();
            mp.release();
            mp = null;
        }
    }
}
```

② 使用 MediaPlayer 类的无参构造函数 MediaPlayer() 创建对象，然后再使用 setDataSource() 方法装载音频文件。这种情况下，MediaPlayer 对象并未真正装载该音频文件，还需要再调用 MediaPlayer 的 prepare() 方法来准备，然后才可以播放该音频文件。

参考代码如下：

```
private MediaPlayer mediaPlayer;
public void playMusic(String path) {
    try {
        //指定要播放的音频文件
        mediaPlayer.setDataSource(path);
        //预加载音频文件
        mediaPlayer.prepare();
        //播放音频
        mediaPlayer.start();
    } catch(Exception e) {
        e.printStackTrace();
    }
}
```

2. 使用 MediaRecorder 录制音频

MediaRecorder 类可以用来设置录制声音的设备、格式和编码器，在录音前需要使用 Environment.getExternalStorageState 判断手机中是否存在 SD 卡，若不存在，则拒绝音频录制。该类的定义如下：

```
java.lang.Object
  ↳ android.media.MediaRecorder
```

MediaRecorder 类常用的方法如表 7-17 所示。

表 7-17 MediaRecorder 常用的方法及说明

序号	方法名称	说明
1	public MediaRecorder( )	无参构造函数,用来创建录音机对象
2	public void prepare( )	准备录音
3	public void release( )	释放录音机对象
4	public void reset( )	重设录音机
5	public void setAudioEncoder(int audio_encoder)	设置音频编码器
6	public void setAudioSource(int audio_source)	设置输入设备
7	public void setOutputFormat(int output_format)	设置输出格式
8	public void setOutputFile(String path)	设置输出文件的路径
9	public void start( )	开始录音
10	public void stop( )	结束录音

7.4.3 使用 SurfaceView 播放视频的步骤

1. 在布局中增加 SurfaceView 组件

播放视频时,视频的显示画面需要放置在 SurfaceView(或者 VideoView)中。Surface-View 类的定义如下:

```
java.lang.Object
  ↳ android.view.View
    ↳ android.view.SurfaceView
```

VideoView 是 SurfaceView 的子类,该类的定义如下:

```
java.lang.Object
  ↳ android.view.View
    ↳ android.view.SurfaceView
      ↳ android.widget.VideoView
```

使用时,可以在 XML 布局文件中进行定义,参考代码如下:

```
<SurfaceView
    android:id = "@ + id/surfaceView"
    android:layout_width = "match_parent"
    android:layout_height = "match_parent"
/>
```

2. 获取 SurfaceView 的 SurfaceHolder

为了能够对 SurfaceView 进行相关操作,需要获取 SurfaceView 的 SurfaceHolder,参考代码如下:

```
surfaceHolder = sv.getHolder( );
```

3. 为 SurfaceHolder 添加回调方法，并重写 surfaceCreated( )

添加回调方法的参考代码：

```
surfaceHolder.addCallback(this);
```

在 SurfaceView 第 1 次创建好后，会调用 surfaceCreated( ) 方法，在其中添加必要的事件代码。

例如，在"悦视播放器"中，使用如下代码实现了当 SurfaceView 准备好时使"播放"按钮可用：

```
@Override
public voidsurfaceCreated(SurfaceHolder holder){
    play.setEnabled(true);
}
```

4. 播放视频

在对视频进行播放等操作时，仍然要使用到 MediaPlayer 对象，调用其相关方法即可实现具体操作。

参考代码如下：

```
//设置要播放的视频路径
mp.setDataSource("/sdcard/Download/2964886.mp4");
//设置将视频画面输出到 SurfaceView
mp.setDisplay(sv.getHolder());
//预加载视频
mp.prepare();
//开始播放
mp.start();
```

5. 暂停和停止视频

```
if(mp.isPlaying()){
    //暂停视频的播放
    mp.pause();
    //停止播放
    //mp.stop();
}
```

7.5 动手实践 7：迷你画板

7.5.1 功能要求

1. 应用程序界面

应用程序的界面如图 7-6 所示。

2. 功能描述

应用程序的首界面如图 7-6a 所示。在界面上方有白色的绘图区，下方有 4 个功能按钮，

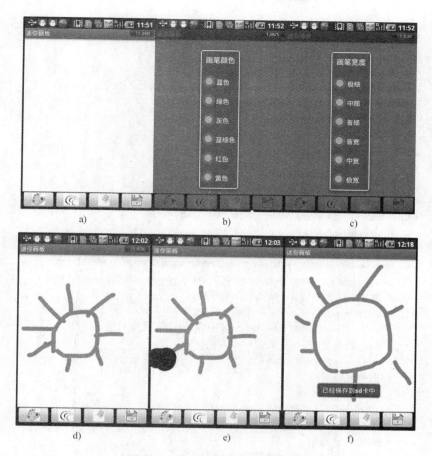

图 7-6 "迷你画板"运行效果图

分别可以让用户设置画笔颜色（默认为红色）、画笔宽度（默认为1sp）、擦除多余内容和保存所绘文件。

当单击第 1 个按钮时，可以弹出如图 7-6b 所示的选择画笔颜色对话框；当单击第 2 个按钮时，可以弹出如图 7-6c 所示的选择画笔宽度对话框；选择好的颜色和宽度可以即时运用在绘图区。

单击"擦除"按钮可将触点所到之处的绘制内容均擦除；单击"保存"按钮可以将绘制的图形保存到 SD 卡中，并设置文件名为"myPicture. png"。

7.5.2 操作提示

首先将程序中需要的图片资源准备好，然后按照以下步骤完成应用程序。

1. 自定义绘图区 DrawView. java

绘图区是自定义组件，其中定义了保存和擦除绘图的方法。参考代码如下：

```java
public class DrawView extends View {
    //屏幕的宽度和宽度
    private int view_width = 0;
    private int view_height = 0;
```

```java
//起始点的坐标值
private float preX, preY;
//绘制路径
private Path path;
//声明画笔
public Paint paint = null;
//声明内存中的图片,该图片将作为缓冲区
Bitmap cacheBitmap = null;
//声明 cacheBitmap 上的 Canvas 对象
Canvas cacheCanvas = null;
//构造函数
public DrawView(Context context, AttributeSet set) {
    super(context, set);
    //以像素为单位获取屏幕的宽度和高度
    view_width = context.getResources().getDisplayMetrics().widthPixels;
    view_height = context.getResources().getDisplayMetrics().heightPixels;
    //创建一个与该 View 相同大小的缓存区
    cacheBitmap = Bitmap.createBitmap(
            view_width, view_height, Config.ARGB_8888);
    //创建画布对象
    cacheCanvas = new Canvas();
    //创建路径对象
    path = new Path();
    //在 cacheCanvas 上绘制 cacheBitmap
    cacheCanvas.setBitmap(cacheBitmap);
    //创建画笔对象
    paint = new Paint(Paint.DITHER_FLAG);
    //设置默认的画笔颜色
    paint.setColor(Color.RED);
    //设置画笔风格
    paint.setStyle(Paint.Style.STROKE);
    paint.setStrokeJoin(Paint.Join.ROUND);
    paint.setStrokeCap(Paint.Cap.ROUND);
    paint.setStrokeWidth(1);
    paint.setAntiAlias(true);
    paint.setDither(true);
}
@Override
public void onDraw(Canvas canvas) {
    //设置背景颜色
    canvas.drawColor(0xFFFFFFFF);
    //采用默认设置创建一个画笔
    Paint bmpPaint = new Paint();
    //绘制 cacheBitmap
    canvas.drawBitmap(cacheBitmap, 0, 0, bmpPaint);
    //绘制路径
    canvas.drawPath(path, paint);
    //保存 canvas 的状态
```

```java
            canvas.save(Canvas.ALL_SAVE_FLAG);
            //恢复 canvas 之前保存的状态
            canvas.restore();
        }
        @Override
        public boolean onTouchEvent(MotionEvent event){
            //获取触摸事件的发生位置
            float x = event.getX();
            float y = event.getY();
            switch(event.getAction()){
            case MotionEvent.ACTION_DOWN:
                //将绘图的起始点移到(x,y)坐标点的位置
                path.moveTo(x,y);
                preX = x;
                preY = y;
                break;
            case MotionEvent.ACTION_MOVE:
                float dx = Math.abs(x - preX);
                float dy = Math.abs(y - preY);
                //判断是否在允许的范围内
                if(dx >=5 || dy >=5){
                    //绘制移动轨迹
                    path.quadTo(preX,preY,(x + preX)/2,(y + preY)/2);
                    preX = x;
                    preY = y;
                }
                break;
            case MotionEvent.ACTION_UP:
                cacheCanvas.drawPath(path,paint);//绘制路径
                path.reset();
                break;
            }
            //重绘 canvas
            invalidate();
            return true;
        }
    }
    //擦除方法
    public void clear(){
        //设置擦除模式
        paint.setXfermode(new PorterDuffXfermode(PorterDuff.Mode.CLEAR));
        //设置笔触的宽度
        paint.setStrokeWidth(30);
    }
    //保存绘制结果的方法
    public void save(){
        try{
            //保存的文件名
            saveBitmap("myPicture");
```

```
            } catch(IOException e) {
                e.printStackTrace();
            }
        }
    //保存绘制好的位图
    public void saveBitmap(String fileName) throws IOException {
        //创建文件对象
        File file = new File("/sdcard/" + fileName + ".png");
        //创建一个新文件
        file.createNewFile();
        //创建一个文件输出流对象
        FileOutputStream fileOS = new FileOutputStream(file);
        //将绘图内容压缩为 PNG 格式输出到输出流对象中
        cacheBitmap.compress(Bitmap.CompressFormat.PNG, 100, fileOS);
        fileOS.flush();      //将缓冲区中的数据全部写出到输出流中
        fileOS.close();      //关闭文件输出流对象
    }
}
```

2. 定义主界面布局 activity_main.xml

使用自定义的 DrawView 和 4 个图片按钮即可生成主界面布局。参考代码如下：

```xml
<?xml version="1.0" encoding="utf-8"?>
<LinearLayout xmlns:android="http://schemas.android.com/apk/res/android"
    android:layout_width="fill_parent"
    android:layout_height="fill_parent"
    android:orientation="vertical" >
<com.book.ex07_1.DrawView
        android:id="@+id/draw_board"
        android:layout_width="match_parent"
        android:layout_height="0dp"
        android:layout_weight="9"/>
<LinearLayout
        android:layout_width="match_parent"
        android:layout_height="0dp"
        android:layout_weight="1" >
<ImageButton
            android:id="@+id/img_color"
            android:layout_width="0dp"
            android:layout_weight="1"
            android:layout_height="wrap_content"
            android:src="@drawable/color_board"
            android:scaleType="fitCenter"
            android:onClick="clickColor"
            />
<ImageButton
            android:layout_width="0dp"
            android:layout_weight="1"
```

```xml
            android:layout_height = "wrap_content"
            android:src = "@drawable/line_width"
            android:scaleType = "fitCenter"
            android:onClick = "clickLine"/>
<ImageButton
            android:layout_width = "0dp"
            android:layout_weight = "1"
            android:layout_height = "wrap_content"
            android:src = "@drawable/eraser"
            android:scaleType = "fitCenter"
            android:onClick = "clickClear"/>
<ImageButton
            android:layout_width = "0dp"
            android:layout_weight = "1"
            android:layout_height = "wrap_content"
            android:src = "@drawable/save"
            android:scaleType = "fitCenter"
            android:onClick = "clickSave"/>
</LinearLayout>
</LinearLayout>
```

3. 设置画笔颜色和画笔宽度对应的布局 color_layout.xml 与 width_layout.xml

画笔颜色和画笔宽度布局中均使用了一组单选按钮，这两个布局非常相似。此处以画笔颜色对应的布局为例，其参考代码如下：

```xml
<?xml version = "1.0" encoding = "utf-8"?>
<LinearLayout xmlns:android = "http://schemas.android.com/apk/res/android"
    android:layout_width = "match_parent"
    android:layout_height = "match_parent"
    android:orientation = "vertical" >
<RadioGroup
        android:id = "@+id/rg_color"
        android:layout_width = "match_parent"
        android:layout_height = "wrap_content" >
<RadioButton
        android:id = "@+id/rdn_blue"
        android:layout_width = "match_parent"
        android:layout_height = "wrap_content"
        android:text = "@string/color_blue"/>
<RadioButton
        android:id = "@+id/rdn_green"
        android:layout_width = "match_parent"
        android:layout_height = "wrap_content"
        android:text = "@string/color_green"/>
<RadioButton
        android:id = "@+id/rdn_gray"
        android:layout_width = "match_parent"
        android:layout_height = "wrap_content"
```

```
                    android:text = "@string/color_gray"/>
        <RadioButton
                    android:id = "@+id/rdn_cyan"
                    android:layout_width = "match_parent"
                    android:layout_height = "wrap_content"
                    android:text = "@string/color_cyan"/>
        <RadioButton
                    android:id = "@+id/rdn_red"
                    android:layout_width = "match_parent"
                    android:layout_height = "wrap_content"
                    android:text = "@string/color_red"/>
        <RadioButton
                    android:id = "@+id/rdn_yellow"
                    android:layout_width = "match_parent"
                    android:layout_height = "wrap_content"
                    android:text = "@string/color_yellow"/>
    </RadioGroup>
</LinearLayout>
```

注意：上述代码中运用了一些字符串资源，需要在 res/values/strings.xml 中进行定义。

4. 定义设置画笔颜色和画笔宽度的活动类 PaintColor.java 与 PaintWidth.java

设置画笔颜色和画笔宽度的活动类的实现也很接近，PaintColor 类的参考代码如下：

```java
public class PaintColor extends Activity implements OnCheckedChangeListener {
    RadioGroup rgColor;
    int color = Color.BLUE;
    @Override
    protected void onCreate(Bundle savedInstanceState) {
        super.onCreate(savedInstanceState);
        setContentView(R.layout.color_layout);
        rgColor = (RadioGroup)findViewById(R.id.rg_color);
        rgColor.setOnCheckedChangeListener(this);
    }
    @Override
    public void onCheckedChanged(RadioGroup group, int checkedId) {
        switch(checkedId) {
        case R.id.rdn_green:
            color = Color.GREEN;
            break;
        case R.id.rdn_gray:
            color = Color.GRAY;
            break;
        case R.id.rdn_cyan:
            color = Color.CYAN;
            break;
        case R.id.rdn_red:
```

```
                color = Color. RED;
                break;
            case R. id. rdn_yellow:
                color = Color. YELLOW;
                break;
            default:
                color = Color. BLUE;
                break;
        }
        Intent i = getIntent( );
        i. putExtra( "color" ,color) ;
        PaintColor. this. setResult( RESULT_OK ,getIntent( ) ) ;
        finish( );
    }
    //屏蔽 Back 键
    @Override
    public boolean onKeyDown( int keyCode ,KeyEvent event) {
        if( keyCode = = KeyEvent. KEYCODE_BACK) {
            return true;
        }
        return super. onKeyDown( keyCode ,event) ;
    }
}
```

5. 在 AndroidManifest. xml 中配置 PaintColor. java 和 PaintWidth. java 类

为了使画笔颜色和画笔宽度的设置均以对话框的形式出现,需要在 AndroidManifest. xml 中进行配置,参考代码如下:

```
< activity
    android:name = "PaintColor"
        android:label = "@string/color_tip"
        android:theme = "@android:style/Theme. Dialog"/>
< activity
    android:name = "PaintWidth"
        android:label = "@string/width_tip"
        android:theme = "@android:style/Theme. Dialog"/>
```

6. 实现 MainActivity 中的功能

MainActivity 中需要调用自定义视图类的方法完成全部画板功能。参考代码如下:

```
public class MainActivity extends Activity {
    DrawView drawBoard;
    final int ColorRequestCode = 1 ,WidthRequestCode = 2 ;
    private int color;
    private int width;
    @Override
    public void onCreate( Bundle savedInstanceState) {
```

```java
        super.onCreate(savedInstanceState);
        setContentView(R.layout.activity_main);
        //获取自定义的绘图视图
        drawBoard = (DrawView)findViewById(R.id.draw_board);
    }
    @Override
    protected void onResume(){
        drawBoard.paint.setXfermode(null);
        super.onResume();
    }
    //单击调色板
    public void clickColor(View v){
        Intent i = new Intent(MainActivity.this,PaintColor.class);
        startActivityForResult(i,ColorRequestCode);
    }
    //单击线宽
    public void clickLine(View v){
        Intent i = new Intent(MainActivity.this,PaintWidth.class);
        startActivityForResult(i,WidthRequestCode);
    }
    //单击橡皮擦
    public void clickClear(View v){
        drawBoard.clear();
    }
    //单击"保存"按钮
    public void clickSave(View v){
        drawBoard.save();
        Toast.makeText(MainActivity.this,"已经保存到SD卡中",
                Toast.LENGTH_SHORT).show();
    }
    //从备选项中获取数据
    @Override
    protected void onActivityResult(int requestCode,int resultCode,Intent data){
        switch(requestCode){
        case ColorRequestCode:
            color = data.getIntExtra("color",Color.BLUE);
            //设置画笔的颜色
            drawBoard.paint.setColor(color);
            break;
        case WidthRequestCode:
            width = data.getIntExtra("width",15);
            //设置笔触的宽度
            drawBoard.paint.setStrokeWidth(width);
            break;
        }
    }
}
```

第 8 章　综合实训：快乐数独

8.1　数独（Sudoku）简介

8.1.1　数独概述

数独是一种源自 18 世纪末的瑞士，后在美国发展，并在日本得以发扬光大的数学智力拼图游戏。拼图是九宫格（即 3 格宽×3 格高）的正方形状，每一格又细分为一个九宫格。在每一个小九宫格中，分别填上 1 至 9 的数字，让整个大九宫格每一列、每一行的数字都不重复。不少教育者认为数独是锻炼脑筋的好方法。

组成数独的基本元素如图 8-1 所示。在标准的数独游戏中共有 81 个"单元格"，它是数独中最小的单元。它们均匀分布在 9×9 的单元格集合中。在图 8-1 中粗黑线划分的区域被称为宫，标准数独中一个宫即是 3×3 的 9 个单元格集合。

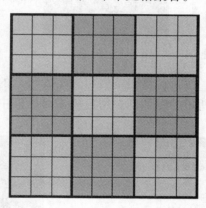

图 8-1　数独基本元素示意图

数独初始盘面中会给出部分数字，被称为已知数；每个空单元格中可以填入的数字，被称为候选数。

8.1.2　数独的游戏规则与技巧

标准数独的规则为：数独中每行、每列及每宫填入数字 1~9 且不能重复。数独解法全是由规则衍生出来的。基本解法分为两类思路，一类为排除法，一类为唯一法。更复杂的解法，最终也会归结到这两大类中。数独直观法解题技巧主要有：唯一解法、基础摒除法、唯余解法等。其中，基础摒除法就是利用 1~9 的数字在每一行、每一列、每一个九宫格都只能出现一次的规则进行解题的方法。基础摒除法可以分为行摒除、列摒除、九宫格摒除。

8.2 项目功能分析

8.2.1 项目的主要功能

如图 8-2 所示，HappySudoku（快乐数独）的主要功能有 4 项，分别是"新游戏"、"继续"游戏、"关于"和"退出"游戏，并且在程序尚未开始"新游戏"之前"继续"按钮不可用。

在用户选择了"新游戏"后，即可弹出如图 8-3 所示的难度选择对话框。当用户选择某种难度等级之后，界面切换到如图 8-4 所示的游戏界面，其中题目从对应等级的题库中随机取出。

图 8-2 HappySudoku 的主要功能

图 8-3 难度选择

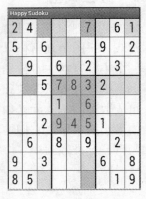
图 8-4 游戏界面

单击首界面中的"关于"按钮，将切换到如图 8-5 所示的"关于"界面。当单击"退出"按钮时，弹出如图 8-6 所示的对话框。用户单击对话框中的"确定"按钮时结束程序运行。

图 8-5 关于

图 8-6 退出

8.2.2 "自定义设置"菜单

在首界面中单击"menu"功能键，即可在界面下方出现"自定义设置"菜单，如图 8-7 所示。单击菜单项，可以切换到设置界面，如图 8-8 所示。

图 8-7　选项菜单

图 8-8　设置界面

8.2.3 "软键盘"与"提示"

在游戏过程中,用户可以按下任何空白单元格,此时会弹出"软键盘"。如果用户设置了"在游戏时给出提示",则软键盘会有提示,如图 8-9 所示。

图 8-9　带提示的软键盘

另外,当用户在单元格中输入错误的数字时,界面播放振动动画,以示提醒。

8.3　准备所需资源

8.3.1　图片(res/drawable-x/)

HappySudoku 项目中用到的图片资源如表 8-1 所示。

表 8-1　HappySudoku 中用到的图片资源说明

序号	资源名称	说　　明
1	bubble_bg.jpg	首界面使用的背景图
2	sudoku.jpg	应用程序图标
3	title.jpg	首界面上方的"Happy Sudoku"

8.3.2 音频（res/raw/）

项目的背景音乐和玩游戏过程中的音乐资源说明如表8-2所示。

表8-2 HappySudoku中用到的音频资源说明

序号	资源名称	说明
1	game.mp3	游戏过程中的音乐
2	main.mp3	背景音乐

8.3.3 数组（res/values/arrays.xml）

各个难度等级的标题用数组来存储，参考代码如下：

```xml
<resources>
<array name="difficulty">
<item>@string/easy_label</item>
<item>@string/medium_label</item>
<item>@string/hard_label</item>
</array>
</resources>
```

8.3.4 颜色（res/values/colors.xml）

绘制数独盘面时需要用到各种不同的颜色资源，参考代码如下：

```xml
<resources>
<!--关于界面和游戏界面的背景颜色 -->
<color name="bg_color">#fff</color>
    <!--绘制数独盘面线条时的颜色 -->
<color name="puzzle_line_color">#000</color>
<color name="puzzle_text_color">#000</color>
<color name="puzzle_hint_0">#64ff0000</color>
<color name="puzzle_hint_1">#6400ff80</color>
<color name="puzzle_hint_2">#2000ff80</color>
<color name="puzzle_selected">#64ff8000</color>
</resources>
```

8.3.5 字符串（res/values/strings.xml）

应用程序中用到了许多字符串资源，参考代码如下：

```xml
<resources>
<string name="app_name">HappySudoku</string>
<string name="continue_label">继续</string>
<string name="new_game_label">新游戏</string>
<string name="about_label">关于</string>
<string name="exit_label">退出</string>
<string name="settings_label">自定义设置</string>
```

```
<string name = "settings_title" >自定义设置</string >
<string name = "settings_shortcut" >s</string >
<string name = "music_title" >音乐</string >
<string name = "music_summary" >播放背景音乐</string >
<string name = "hints_title" >提示</string >
<string name = "hints_summary" >在游戏时给出提示</string >
<string name = "new_game_title" >难度选择</string >
<string name = "easy_label" >容易</string >
<string name = "medium_label" >中等</string >
<string name = "hard_label" >困难</string >
<string name = "game_title" >游戏</string >
<string name = "no_moves_label" >对不起,此处已无法填充</string >
<string name = "keypad_title" >软键盘</string >
<string name = "about_title" >关于</string >
<string name = "about_text" >关于的显示内容...</string >
</resources >
```

8.3.6 动画（res/anim/cycle.xml 和 shake.xml）

当用户在单元格中输入错误的数字时,播放振动动画,cycle.xml 的参考代码如下:

```
<cycleInterpolator
    xmlns:android = "http://schemas.android.com/apk/res/android"
    android:cycles = "7"/>
```

shake.xml 的参考代码如下:

```
<translate
    xmlns:android = "http://schemas.android.com/apk/res/android"
    android:fromXDelta = "0"
    android:toXDelta = "10"
    android:duration = "1000"
    android:interpolator = "@anim/cycle"/>
```

8.4 界面设计

8.4.1 首界面（res/layout/activity_main.xml）

首界面使用垂直线性布局实现,其中有 1 个 ImageView 和 4 个按钮。参考代码如下:

```
<LinearLayout
    xmlns:android = "http://schemas.android.com/apk/res/android"
    android:background = "@drawable/bubble_bg"
    android:layout_height = "match_parent"
    android:layout_width = "match_parent"
    android:padding = "30dip"
    android:layout_gravity = "center"
```

```xml
        android:orientation = "vertical" >
<ImageView
        android:id = "@+id/imageView1"
        android:layout_width = "wrap_content"
        android:layout_height = "wrap_content"
        android:src = "@drawable/title"
        android:padding = "10dp"/>
<Button
        android:id = "@+id/new_button"
        android:layout_width = "match_parent"
        android:layout_height = "wrap_content"
        android:text = "@string/new_game_label"/>
<Button
        android:id = "@+id/continue_button"
        android:layout_width = "match_parent"
        android:layout_height = "wrap_content"
        android:text = "@string/continue_label"
        android:enabled = "false"/>
其余按钮可参照第1个完成...
</LinearLayout>
```

8.4.2 游戏界面（SudokuView.java）

游戏界面为自己绘制数独盘面的 View，参考代码如下：

```java
public class SudokuView extends View {
    private static final String SELX = "selX";
    private static final String SELY = "selY";
    private static final String VIEW_STATE = "viewState";
    //序列号
    private static final int ID = 42;
    //单元格的宽度和高度
    private float width, height;
    //选中的单元格坐标
    private int selX, selY;
    //选中的矩形
    private final Rect selRect = new Rect();
    //使用 View 的 Activity 类
    private final NewGameActivity game;
    public SudokuView(Context context) {
        super(context);
        this.game = (NewGameActivity) context;
        setFocusable(true);
        setFocusableInTouchMode(true);
        setId(ID);
    }
    @Override
    protected Parcelable onSaveInstanceState() {
```

```java
        Parcelable p = super.onSaveInstanceState();
        Bundle bundle = new Bundle();
        bundle.putInt(SELX, selX);
        bundle.putInt(SELY, selY);
        bundle.putParcelable(VIEW_STATE, p);
        return bundle;
}
@Override
protected void onRestoreInstanceState(Parcelable state) {
        Bundle bundle = (Bundle) state;
        select(bundle.getInt(SELX), bundle.getInt(SELY));
        super.onRestoreInstanceState(bundle.getParcelable(VIEW_STATE));
        return;
}
@Override
protected void onSizeChanged(int w, int h, int oldw, int oldh) {
        width = w/9f;
        height = h/9f;
        getRect(selX, selY, selRect);
        super.onSizeChanged(w, h, oldw, oldh);
}


@Override
protected void onDraw(Canvas canvas) {
        //绘制盘面背景
        Paint background = new Paint();
        background.setColor(getResources().getColor(R.color.bg_color));
        canvas.drawRect(0, 0, getWidth(), getHeight(), background);
        //准备粗画笔
        Paint dark = new Paint();
        dark.setColor(getResources().getColor(R.color.puzzle_line_color));
        dark.setStrokeWidth(3);
        //准备细画笔
        Paint light = new Paint();
        light.setColor(getResources().getColor(R.color.puzzle_line_color));
        light.setStrokeWidth(1);
        //用细画笔分别绘制垂直方向和水平方向的单元格线
        for(int i = 1; i <= 9; i++) {
            canvas.drawLine(0, i * height, getWidth(), i * height, light);
            canvas.drawLine(i * width, 0, i * width, getHeight(), light);
        }
        //用粗画笔绘制宫线
        for(int i = 1; i <= 9; i++) {
            if(i%3 != 0)
                continue;
            canvas.drawLine(0, i * height, getWidth(), i * height, dark);
```

```
            canvas.drawLine(i*width,0,i*width,getHeight(),dark);
}
//设置绘制数字的画笔
Paint textPaint = new Paint(Paint.ANTI_ALIAS_FLAG);
textPaint.setColor(getResources().getColor(R.color.puzzle_text_color));
textPaint.setStyle(Style.FILL);
textPaint.setTextSize(height*0.75f);
textPaint.setTextScaleX(width/height);
textPaint.setTextAlign(Paint.Align.CENTER);
//将数字写在单元格的中间
FontMetrics fm = textPaint.getFontMetrics();
//将数字写在水平方向的中间位置
float x = width/2;
//将数字写在竖直方向的中间位置
float y = height/2 - (fm.ascent + fm.descent)/2;
//在指定位置绘制数字
for(int i = 0;i < 9;i++){
    for(int j = 0;j < 9;j++){
        canvas.drawText(this.game.getTileString(i,j),i*width + x,j*height + y,textPaint);
    }
}
if(SettingActivity.getHints(getContext())){
    //如果单元格内可以填的数字的数量小于3,则将单元格加深颜色显示
    Paint hint = new Paint();
    int c[] = {
        getResources().getColor(R.color.puzzle_hint_0),
        getResources().getColor(R.color.puzzle_hint_1),
        getResources().getColor(R.color.puzzle_hint_2)};
    Rect r = new Rect();
    for(int i = 0;i < 9;i++){
        for(int j = 0;j < 9;j++){
            intmovesleft = 9 - game.getUsedTiles(i,j).length;
            if(movesleft < c.length){
                getRect(i,j,r);
                hint.setColor(c[movesleft]);
                canvas.drawRect(r,hint);
            }
        }
    }
}
//设置选中单元格的画笔,并绘制选中的单元格
Paint selected = new Paint();
selected.setColor(getResources().getColor(R.color.puzzle_selected));
canvas.drawRect(selRect,selected);
}
```

```java
@Override
public booleanonTouchEvent(MotionEvent event) {
    if(event.getAction() != MotionEvent.ACTION_DOWN)
        return super.onTouchEvent(event);
    //取得当前单击的单元格位置(x,y)
    int x = (int)(event.getX()/width);
    int y = (int)(event.getY()/height);
    //选择当前单元格
    select(x,y);
    //如果单元格里没有数,显示小键盘
    if(game.getTileString(x,y) == "") {
        game.showKeypadOrError(selX,selY);
    }
    return true;
}
//按键事件
@Override
public booleanonKeyDown(int keyCode,KeyEvent event) {
    switch(keyCode) {
    //方向键执行选择单元格的操作
    caseKeyEvent.KEYCODE_DPAD_UP:
        select(selX,selY - 1);
        break;
    caseKeyEvent.KEYCODE_DPAD_DOWN:
        select(selX,selY + 1);
        break;
    caseKeyEvent.KEYCODE_DPAD_LEFT:
        select(selX - 1,selY);
        break;
    caseKeyEvent.KEYCODE_DPAD_RIGHT:
        select(selX + 1,selY);
        break;
    //数字键执行填数的操作
    caseKeyEvent.KEYCODE_0:
    caseKeyEvent.KEYCODE_SPACE:   setSelectedTile(0);break;
    caseKeyEvent.KEYCODE_1:       setSelectedTile(1);break;
    caseKeyEvent.KEYCODE_2:       setSelectedTile(2);break;
    caseKeyEvent.KEYCODE_3:       setSelectedTile(3);break;
    caseKeyEvent.KEYCODE_4:       setSelectedTile(4);break;
    caseKeyEvent.KEYCODE_5:       setSelectedTile(5);break;
    caseKeyEvent.KEYCODE_6:       setSelectedTile(6);break;
    caseKeyEvent.KEYCODE_7:       setSelectedTile(7);break;
    caseKeyEvent.KEYCODE_8:       setSelectedTile(8);break;
    caseKeyEvent.KEYCODE_9:       setSelectedTile(9);break;
    //按回车键或中间键则显示软键盘或给出错误提示
    caseKeyEvent.KEYCODE_ENTER:
    caseKeyEvent.KEYCODE_DPAD_CENTER:
```

```
                game.showKeypadOrError(selX,selY);
                break;
            default:
                return super.onKeyDown(keyCode,event);
        }
        return true;
    }
    //为选中单元格填充数字
    public void setSelectedTile(int tile){
        //如果值正确则重新绘制数独盘面,改变提示数字
        if(game.setTileIfValid(selX,selY,tile)){
            invalidate();//
        } else {
            //如果值不正确则播放振动动画
            startAnimation(AnimationUtils.loadAnimation(game,R.anim.shake));
        }
    }
    //根据坐标位置选择单元格
    private void select(int x,int y){
        invalidate(selRect);
        selX = Math.min(Math.max(x,0),8);
        selY = Math.min(Math.max(y,0),8);
        //获取选中的单元格 rect
        getRect(selX,selY,selRect);
        invalidate(selRect);
    }
    //根据坐标位置获取选中的单元格 rect
    private void getRect(int x,int y,Rect rect){
        rect.set((int)(x*width),(int)(y*height),(int)(x*width+width),(int)(y*height+height));
    }
}
```

8.4.3 设置界面(res/xml/settings.xml)

用 PreferenceScreen 类中的常见设置组件可以高效地生成设置界面,参考代码如下:

```
<PreferenceScreen
    xmlns:android="http://schemas.android.com/apk/res/android">
<CheckBoxPreference
        android:key="music"
        android:title="@string/music_title"
        android:summary="@string/music_summary"
        android:defaultValue="true"/>
<CheckBoxPreference
        android:key="hints"
        android:title="@string/hints_title"
        android:summary="@string/hints_summary"
```

```
        android:defaultValue = "true"/>
</PreferenceScreen>
```

8.4.4 软键盘界面（res/layout/keypad.xml）

软键盘显示 9 个数字，使用表格布局实现，参考代码如下：

```
<TableLayout
    xmlns:android = "http://schemas.android.com/apk/res/android"
    android:id = "@+id/keypad"
    android:orientation = "vertical"
    android:layout_width = "wrap_content"
    android:layout_height = "wrap_content"
    android:stretchColumns = " * ">
<TableRow>
<Button android:id = "@+id/keypad_1"
        android:text = "1" >
</Button>
<Button android:id = "@+id/keypad_2"
        android:text = "2" >
</Button>
<Button android:id = "@+id/keypad_3"
        android:text = "3" >
</Button>
</TableRow>
其余两行类似,不再赘述...
</TableLayout>
```

8.4.5 菜单界面（res/menu/menu.xml）

菜单界面比较简单，参考代码如如下：

```
<menu xmlns:android = "http://schemas.android.com/apk/res/android">
<item android:id = "@+id/settings"
      android:title = "@string/settings_label"
      android:alphabeticShortcut = "@string/settings_shortcut"/>
</menu>
```

关于界面（res/layout/about.xml）更加简单，使用 ScrollView 作根结点，1 个 TextView 为其子组件即可。

8.5 数据库设计

8.5.1 定义数据库常量类（Constants.java）

在这个类中定义了数据表和表中各列的名称，参考代码如下：

```java
public interface Constants extends BaseColumns {
    public static final String TABLE_NAME = "tb_puzzle";
    public static final String TIMU = "tm";
    public static final String NANDU = "nd";
}
```

8.5.2 定义数据库辅助类（DBHelper.java）

在这个类中创建了数据库 puzzle.db 和数据表，并为数据表中 3 种难度等级各添加了 2 个题目，参考代码如下：

```java
public class DBHelper extends SQLiteOpenHelper {
    //定义数据库的名称
    static final String DB_NAME = "puzzle.db";
    static final int DB_VERSION = 1;
    public DBHelper(Context context) {
        super(context, DB_NAME, null, DB_VERSION);
    }
    @Override
    public void onCreate(SQLiteDatabase db) {
        //创建数据表
        String sql = "CREATE TABLE" + TABLE_NAME + "("
                + _ID + " INTEGER PRIMARY KEY AUTOINCREMENT,"
                + TIMU + " TEXT NOT NULL,"
                + NANDU + " INTEGER);";
        db.execSQL(sql);
        //为数据库添加题目
        addData(db);
    }
    @Override
    public void onUpgrade(SQLiteDatabase db, int oldVersion, int newVersion) {
        String sql = "DROP TABLE IF EXISTS" + TABLE_NAME;
        db.execSQL(sql);
        onCreate(db);
    }
    //为数据库添加题目
    private void addData(SQLiteDatabase db) {
        //添加容易级别的题目
        ContentValues v = new ContentValues();
        v.put(TIMU, "360000000004230800000004200" +
                "070460003820000014500013020" +
                "001900000007048300000000045");
        v.put(NANDU, 0);
        db.insertOrThrow(TABLE_NAME, null, v);
        v.put(TIMU, "240000061506000902090602030" +
                "005783200000106000002945100" +
```

```
                    "060809020903000608850000019");
            v.put(NANDU,0);
            db.insertOrThrow(TABLE_NAME,null,v);
            //添加中等级别的题目
            v.put(TIMU,"650000070000506000014000005" +
                    "007009000002314700000700800" +
                    "500000630000201000030000097");
            v.put(NANDU,1);
            db.insertOrThrow(TABLE_NAME,null,v);
            v.put(TIMU,"930000074510000083002805100" +
                    "004207900000010000007309400" +
                    "001703600260000059740000018");
            v.put(NANDU,1);
            db.insertOrThrow(TABLE_NAME,null,v);
            //添加高难度级别的题目
            v.put(TIMU,"009000000080605020501078000" +
                    "000000700706040102004000000" +
                    "000720903090301080000000600");
            v.put(NANDU,2);
            db.insertOrThrow(TABLE_NAME,null,v);
            v.put(TIMU,"000000000010000584062400070" +
                    "004050302000010000208030600" +
                    "080003960496000020000000000");
            v.put(NANDU,2);
            db.insertOrThrow(TABLE_NAME,null,v);
        }
    }
```

8.6 功能实现与完善

8.6.1 首界面中按钮与菜单的功能（MainActivity.java）

首界面中主要实现按钮的单击和选项菜单的功能，在以下参考代码中用到了若干其他类（加粗显示部分），读者可以根据后文的提示创建这些类。

```java
public class MainActivity extends Activity implements OnClickListener {
    private Button continueButton,newButton,aboutButton,exitButton;
    @Override
    public void onCreate(Bundle savedInstanceState) {
        super.onCreate(savedInstanceState);
        setContentView(R.layout.activity_main);
        //为按钮添加单击监听
        continueButton = (Button)findViewById(R.id.continue_button);
        continueButton.setOnClickListener(this);
        newButton = (Button)findViewById(R.id.new_button);
        newButton.setOnClickListener(this);
```

```java
            aboutButton = (Button)findViewById(R.id.about_button);
            aboutButton.setOnClickListener(this);
            exitButton = (Button)findViewById(R.id.exit_button);
            exitButton.setOnClickListener(this);
    }
    @Override
    protected void onResume(){
        super.onResume();
        Music.play(this,R.raw.main);
        continueButton.setEnabled(true);
    }
    @Override
    protected void onPause(){
        super.onPause();
        Music.stop(this);
    }
    //实现单击监听
    public void onClick(View v){
        switch(v.getId()){
        case R.id.continue_button:
            startGame(NewGameActivity.DIFFICULTY_CONTINUE);
            break;
        case R.id.about_button:
            Intent i = new Intent(this,AboutActivity.class);
            startActivity(i);
            break;
        case R.id.new_button:
            //显示难度等级对话框
            openDifficultyDialog();
            continueButton.setEnabled(true);
            break;
            //单击"退出"按钮时,显示对话框
        case R.id.exit_button:
            showAlertDialog();
            break;
        }
    }
    //显示对话框
    private void showAlertDialog(){
        AlertDialog.Builder ad = new AlertDialog.Builder(MainActivity.this);
        ad.setIcon(android.R.drawable.ic_dialog_alert)
        .setTitle(R.string.exit_label)
        .setMessage("您确定要退出吗?")
        .setPositiveButton(android.R.string.ok,
            new android.content.DialogInterface.OnClickListener(){
                public void onClick(DialogInterface dialog,int which){
                    //单击"OK"按钮则退出应用程序
                    finish();
```

```java
            }
        })
        .setNegativeButton(android.R.string.cancel,
            new android.content.DialogInterface.OnClickListener(){
                public void onClick(DialogInterface dialog,int which){
                    //单击"cancel"按钮,不做任何事情
                }})
        .create()
        .show();
}
@Override
public boolean onCreateOptionsMenu(Menu menu){
    //创建选项菜单
    super.onCreateOptionsMenu(menu);
    MenuInflater inflater = getMenuInflater();
    inflater.inflate(R.menu.menu,menu);
    return true;
}
@Override
public boolean onOptionsItemSelected(MenuItem item){
    //选择了菜单后的功能
    switch(item.getItemId()){
        case R.id.settings:
            startActivity(new Intent(this,SettingActivity.class));
            return true;
    }
    return false;
}

//选择难易程度的对话框
private void openDifficultyDialog(){
    new AlertDialog.Builder(this)
        .setTitle(R.string.new_game_title)
        .setItems(R.array.difficulty,
            new DialogInterface.OnClickListener(){
                public void onClick(DialogInterface dialoginterface,int i){
//根据选择的难度等级确定开始哪个游戏
    startGame(i);
                }
            })
        .show();
}
//根据选择的难度等级确定开始哪个游戏
private void startGame(int i){
    Intent intent = new Intent(MainActivity.this,NewGameActivity.class);
    intent.putExtra(NewGameActivity.KEY_DIFFICULTY,i);
    startActivity(intent);
}
}
```

8.6.2 "设置"的实现

1. 设置功能（SettingsActivity.java）

使用 SharedPreferences 存取用户的各项设置值，参考代码如下：

```java
public class SettingActivity extends PreferenceActivity {
    //设置的标签和设置状态
    private static final String OPT_MUSIC = "music";
    private static final boolean OPT_MUSIC_DEF = true;
    private static final String OPT_HINTS = "hints";
    private static final boolean OPT_HINTS_DEF = true;
    @Override
    protected void onCreate(Bundle savedInstanceState) {
        //根据 XML 文件创建设置界面
        super.onCreate(savedInstanceState);
        addPreferencesFromResource(R.xml.settings);
    }
    //获取当前的"播放背景音乐"值,默认为播放
    public static boolean getMusic(Context context) {
        return PreferenceManager.getDefaultSharedPreferences(context)
                .getBoolean(OPT_MUSIC, OPT_MUSIC_DEF);
    }
    //获取当前的"给出提示"值,默认为给提示
    public static boolean getHints(Context context) {
        return PreferenceManager.getDefaultSharedPreferences(context)
                .getBoolean(OPT_HINTS, OPT_HINTS_DEF);
    }
}
```

2. 控制音频（Music.java）

```java
public class Music {
    //声明音频播放器对象
    private static MediaPlayer mp = null;
    //停止旧音乐,播放新音乐
    public static void play(Context context, int resource) {
        stop(context);
        //只有用户设置"播放音乐"时,才可播放
        if(SettingActivity.getMusic(context)) {
            mp = MediaPlayer.create(context, resource);
            //循环播放
            mp.setLooping(true);
            mp.start();
        }
    }
    //停止音乐
    public static void stop(Context context) {
        if(mp != null) {
```

```java
                mp.stop();
                mp.release();
                mp = null;
            }
        }
    }
```

3. 保存用户的设置（UserOptions.java）

```java
public class UserOptions {
    private static final String SUDOKU_OPTIONS = MainActivity.class.getName();
    private static final String OPT_MUSIC = "music";
    private static final boolean OPT_MUSIC_DEF = true;
    private static final String OPT_HINTS = "hints";
    private static final boolean OPT_HINTS_DEF = true;
    private static SharedPreferences getSudokuPreferences(
            Context context) {
        return context.getSharedPreferences(SUDOKU_OPTIONS,
                Context.MODE_PRIVATE);
    }
    public static boolean getMusic(Context context) {
        return getSudokuPreferences(context).getBoolean(
                OPT_MUSIC, OPT_MUSIC_DEF);
    }
    public static boolean getHints(Context context) {
        return getSudokuPreferences(context).getBoolean(
                OPT_HINTS, OPT_HINTS_DEF);
    }
    public static boolean putMusic(Context context, boolean value) {
        return getSudokuPreferences(context)
                .edit()
                .putBoolean(OPT_MUSIC, value)
                .commit();
    }
    public static boolean putHints(Context context, boolean value) {
        return getSudokuPreferences(context)
                .edit()
                .putBoolean(OPT_HINTS, value)
                .commit();
    }
}
```

8.6.3 "新游戏"与"继续"功能（NewGameActivity.java）

1. 游戏功能

游戏时，首先要从数据库中读取用户选择难度等级的题目，在用户单击空白单元格后显示软键盘，与此同时，计算可用的数据有哪些，给出提示。参考代码如下：

```java
public class NewGameActivity extends Activity {
    public static final String KEY_DIFFICULTY = "difficulty";
    private static final String PREF_PUZZLE = "puzzle";
    public static final int DIFFICULTY_EASY = 0;
    public static final int DIFFICULTY_MEDIUM = 1;
    public static final int DIFFICULTY_HARD = 2;
    protected static final int DIFFICULTY_CONTINUE = -1;
    private int puzzle[] = new int[9*9];
    private DBHelper db;
    private SudokuView puzzleView;
    @Override
    protected void onCreate(Bundle savedInstanceState) {
        super.onCreate(savedInstanceState);
        db = new DBHelper(this);
        int diff = getIntent().getIntExtra(KEY_DIFFICULTY, DIFFICULTY_EASY);
        puzzle = getPuzzle(diff);
        calculateUsedTiles();
        puzzleView = new SudokuView(this);
        setContentView(puzzleView);
        puzzleView.requestFocus();
        //继续游戏则将 DIFFICULTY_CONTINUE 存到 KEY_DIFFICULTY 中,以便下次依旧可以继续
        getIntent().putExtra(KEY_DIFFICULTY, DIFFICULTY_CONTINUE);
    }
    @Override
    protected void onResume() {
        super.onResume();
        Music.play(this, R.raw.game);
    }
    @Override
    protected void onPause() {
        super.onPause();
        //保存当前的题目
        getPreferences(MODE_PRIVATE).edit().putString(PREF_PUZZLE,
                toPuzzleString(puzzle)).commit();
    }
    //根据难度等级取题
    private int[] getPuzzle(int diff) {
        String puz = null;
        if(diff == DIFFICULTY_CONTINUE) {
            puz = getPreferences(MODE_PRIVATE).getString(PREF_PUZZLE, null);
        }
        else{
            String[] FROM_DIFF = {_ID, TIMU, NANDU};
            SQLiteDatabase db_t = db.getReadableDatabase();
            String selection = NANDU + " = " + diff;
            Cursor cursor = db_t.query(TABLE_NAME, FROM_DIFF, selection, null, null, null, null);
            int tmCount = cursor.getCount();
```

```java
            if( tmCount > 1 ) {
                int tmId = new Random( ).nextInt( tmCount );
                for( int i = 0;i < = tmId;i + + ) {
                    cursor.moveToNext( );
                }
            }
            int tm = cursor.getColumnIndex( TIMU );
            puz = cursor.getString( tm );
            cursor.close( );
        }
        return fromPuzzleString( puz );
    }
    //将整形数组转换为字符串
    static private String toPuzzleString( int[ ] puz ) {
        StringBuilder buf = new StringBuilder( );
        for( int element : puz ) {
            buf.append( element );
        }
        return buf.toString( );
    }
    //将数独题目从字符串转换为整型数组
    static protected int[ ] fromPuzzleString( String string ) {
        int[ ] puz = new int[ string.length( ) ];
        for( int i = 0;i < puz.length;i + + ) {
            puz[ i ] = string.charAt( i ) - '0';
        }
        return puz;
    }
    //取得单元格内的数字值
    private int getTile( int x,int y ) {
        return puzzle[ y * 9 + x ];
    }

    //改变单元格内的数字值
    private void setTile( int x,int y,int value ) {
        puzzle[ y * 9 + x ] = value;
    }
    //取得单元格内数字的字符串值
    protected String getTileString( int x,int y ) {
        int v = getTile( x,y );
        if( v = = 0 )
            return " ";
        else
            return String.valueOf( v );
    }
    //只有输入的数字合法时才将数字写到单元格
    protected boolean setTileIfValid( int x,int y,int value ) {
        int tiles[ ] = getUsedTiles( x,y );
```

```java
            if( value !=0) {
                for( int tile : tiles) {
                    if( tile == value)
                        return false;
                }
            }
            setTile(x,y,value);
            calculateUsedTiles();
            return true;
}
//显示软键盘或错误提示
protected void showKeypadOrError( int x,int y) {
    int tiles[ ] = getUsedTiles(x,y);
    if( tiles. length ==9) {
        Toast toast = Toast. makeText( this,
                R. string. no_moves_label,Toast. LENGTH_SHORT);
        toast. setGravity( Gravity. CENTER ,0,0);
        toast. show();
    } else {
        Dialog v = new Keypad( this,tiles,puzzleView);
        v. show();
    }
}
//声明某单元格中不能再使用的数字数组
private final int used[ ][ ][ ] = new int[9][9][ ];
//取得单元格中不能再使用的数字
protected int[ ] getUsedTiles( int x,int y) {
    return used[x][y];
}
//计算某单元格中不能再使用的数字,并保存在数组中
private void calculateUsedTiles() {
    for( int x =0;x <9;x ++) {
        for( int y =0;y <9;y ++ ) {
            used[x][y] = calculateUsedTiles(x,y);
        }
    }
}
//计算当前单元格处不能再填哪些数字
private int[ ] calculateUsedTiles( int x,int y) {
    int c[ ] = new int[9];
    //本行已经使用了哪些数字
    for( int i =0;i <9;i ++) {
        if( i == y)
            continue;
        int t = getTile(x,i);
        if( t!=0)
            c[t-1] = t;
    }
```

```java
        //本列已经使用了哪些数字
        for(int i=0;i<9;i++){
            if(i==x)
                continue;
            int t = getTile(i,y);
            if(t!=0)
                c[t-1] = t;
        }
        //单元格所在宫中已经使用过的数字
        int startx = (x/3)*3;
        int starty = (y/3)*3;
        for(int i=startx;i<startx+3;i++){
            for(int j=starty;j<starty+3;j++){
                if(i==x && j==y)
                    continue;
                int t = getTile(i,j);
                if(t!=0)
                    c[t-1] = t;
            }
        }
        //统计使用过的数字,保存在c1中
        int nused = 0;
        for(int t:c){
            if(t!=0)
                nused++;
        }
        int c1[] = new int[nused];
        nused = 0;
        for(int t:c){
            if(t!=0)
                c1[nused++] = t;
        }
        return c1;
    }
}
```

2. 软键盘的实现（Keypad. java）

```java
public class Keypad extends Dialog {
    protected static final String TAG = "Sudoku";
    private final View keys[] = new View[9];
    private View keypad;
    private final int useds[];
    private final SudokuView puzzleView;
    public Keypad(Context context, int useds[], SudokuView puzzleView){
        super(context);
        this.useds = useds;
        this.puzzleView = puzzleView;
```

```java
    }
    @Override
    protected void onCreate(Bundle savedInstanceState) {
        super.onCreate(savedInstanceState);
        setTitle(R.string.keypad_title);
        setContentView(R.layout.keypad);
        findViews();
        for(int element : useds) {
            if(element != 0)
                keys[element - 1].setVisibility(View.INVISIBLE);
        }
        setListeners();
    }
    @Override
    public boolean onKeyDown(int keyCode, KeyEvent event) {
        int tile = 0;
        switch(keyCode) {
        case KeyEvent.KEYCODE_0:
        case KeyEvent.KEYCODE_SPACE: tile = 0; break;
        case KeyEvent.KEYCODE_1:     tile = 1; break;
        case KeyEvent.KEYCODE_2:     tile = 2; break;
        case KeyEvent.KEYCODE_3:     tile = 3; break;
        case KeyEvent.KEYCODE_4:     tile = 4; break;
        case KeyEvent.KEYCODE_5:     tile = 5; break;
        case KeyEvent.KEYCODE_6:     tile = 6; break;
        case KeyEvent.KEYCODE_7:     tile = 7; break;
        case KeyEvent.KEYCODE_8:     tile = 8; break;
        case KeyEvent.KEYCODE_9:     tile = 9; break;
        default:
            return super.onKeyDown(keyCode, event);
        }
        if(isValid(tile)) {
            returnResult(tile);
        }
        return true;
    }
}
//填数
/** Return the chosen tile to the caller */
private void returnResult(int tile) {
    puzzleView.setSelectedTile(tile);
    dismiss();
}
//判断是否合法
private boolean isValid(int tile) {
    for(int t : useds) {
        if(tile == t)
            return false;
    }
```

```
            return true;
    }

    private void findViews() {
        keypad = findViewById(R.id.keypad);
        keys[0] = findViewById(R.id.keypad_1);
        keys[1] = findViewById(R.id.keypad_2);
        keys[2] = findViewById(R.id.keypad_3);
        keys[3] = findViewById(R.id.keypad_4);
        keys[4] = findViewById(R.id.keypad_5);
        keys[5] = findViewById(R.id.keypad_6);
        keys[6] = findViewById(R.id.keypad_7);
        keys[7] = findViewById(R.id.keypad_8);
        keys[8] = findViewById(R.id.keypad_9);
    }

    private void setListeners() {
        for(int i = 0; i < keys.length; i ++) {
            final int t = i + 1;
            keys[i].setOnClickListener(new View.OnClickListener() {
                public void onClick(View v) {
                    returnResult(t);
                }});
        }
        keypad.setOnClickListener(new View.OnClickListener() {
            public void onClick(View v) {
                returnResult(0);
            }});
    }
}
```

8.6.4 "关于"功能（AboutActivity.java）

```
public class AboutActivity extends Activity {
    @Override
    protected void onCreate(Bundle savedInstanceState) {
        super.onCreate(savedInstanceState);
        setContentView(R.layout.about);
    }
}
```

8.6.5 修改配置文件（AndroidManifest.xml）

```
<?xml version = "1.0" encoding = "utf-8"?>
<manifest xmlns:android = "http://schemas.android.com/apk/res/android"
    package = "com.book.happysudoku"
```

```xml
            android:versionCode = "1"
            android:versionName = "1.0" >
    <uses-sdk android:minSdkVersion = "8"/>
    <application
            android:icon = "@drawable/sudoku"
            android:label = "@string/app_name" >
        <activity
                android:label = "@string/app_name"
                android:name = ".MainActivity" >
            <intent-filter >
            <action android:name = "android.intent.action.MAIN"/>
            <category android:name = "android.intent.category.LAUNCHER"/>
            </intent-filter >
        </activity >
        <activity
                android:name = "AboutActivity"
                android:theme = "@android:style/Theme.Dialog"
                android:label = "@string/about_label"/>
        <activity
                android:name = "SettingActivity"/>
        <activity
                android:name = "NewGameActivity"/>
    </application >
</manifest >
```

8.6.6 项目的完善方向

本项目实现了数独游戏的基本功能，开发者还可以从以下几个方面对应用程序做出进一步的完善。

1. 更改数字的输入模式

移动设备支持多种方式的信息输入，例如利用各种传感器输入信息，或者通过手势输入信息等。感兴趣的开发者可以使用其他技术实现数字的输入模式。

2. 允许用户擦除输入的数字

游戏中应该允许用户反悔，对其所填入的数字进行更改，开发者可以在本项目的基础上对此功能进行深入完善。

3. 允许用户选择背景音乐

现有项目的背景音乐是指定的音乐文件，开发者还可以增加相关功能，使用户可以从其音乐库中选择背景音乐。

4. 为游戏增加开场动画

许多游戏在开始时，都有开场动画，或者用来显示开发者信息，或者用开场动画的形式加载相关数据等，数独游戏也可以这样华丽开场。

APP时代，想要让自己的应用经久不衰，需要开发者们不仅充分考虑应用的功能，更要考虑到用户的感受，关爱用户。